GENE ALEXANDRESCU

ULTIMATE SECRETS OF REALITY

FROM PARTICLES TO UNIVERSES,

INFINITE REALMS UNFOLD BEYOND SIGHT

ULTIMATE SECRETS OF REALITY

FROM PARTICLES TO UNIVERSES, INFINITE REALMS UNFOLD BEYOND SIGHT

GENE ALEXANDRESCU
CONVERGETICS SCIENCE CENTER

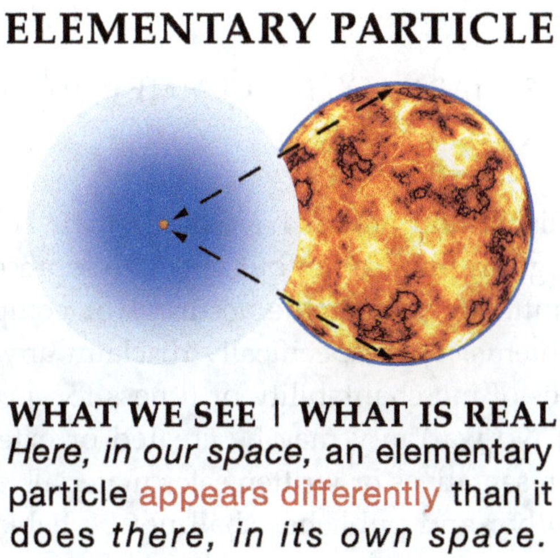

Figure 1: Elementary particle.
Credit: CSC CC BY-NC-ND 4.0.

CONVERGETICS, PUBLISHERS

LIMIT OF LIABILITY/DISCLAIMER OF WARRANTY

Published by Convergetics, Publishers, Montreal, Canada.

NOTE TO LIBRARIANS

A cataloguing record for this book is available from Library and Archives Canada at www.collectionscanada.ca/amicus/index-e.html

10 9 8 7 6 5 4 3 2 1

Ultimate secrets of reality

From particles to universes, infinite realms unfold beyond sight

Gene Alexandrescu

Book ISBN: 978-1-0697151-0-4
Hardcover ISBN: 978-1-0697151-1-1
Digital ISBN: 978-1-0697151-3-5

In September 1908, Hermann Minkowski introduced the postulate of the absolute world with an infinite number of spaces that transformed our understanding of reality:

> "... *we would then have in the world no more* the space, *but an infinite number of spaces analogously as there is an infinite number of planes in three-dimensional space. Three-dimensional geometry becomes a chapter in four-dimensional physics.*"

Minkowski's phrase describes a paradigm shift in human understanding of reality, transitioning from a single-space universe to the paradigm of reality with infinite spaces.

This shift unveils the ultimate secrets shaping our reality, from tiny particles to vast universes.

Minkowski's discovery was so transformative that it remained misunderstood by scientists for 117 years—until now!

—Author

This book is dedicated to three visionary scientists: Hermann Minkowski (1864 - 1909), Rafael Bombelli (1526 - 1572), and Samuel Warren Carey (1911 - 2002).

Each made groundbreaking contributions in their fields, with ideas that were—and even today remain—far ahead of their time. Unfortunately, they were not understood until now and passed away without full recognition. Consequently, the scientific community did not advance their work.

Had their insights been understood earlier, they could have significantly influenced our understanding of reality.

Minkowski's discovery that reality consists of infinite spaces is central to this book. Even today, his idea remains a profound mystery for both the scientific community and the world. Regrettably, Einstein grasped only a fraction of Minkowski's theory, leading us to view the universe as a singular, continuous spacetime, while neglecting the complexity of Minkowski's theory.

In mathematics, Bombelli introduced the concept of imaginary numbers—square roots of negative numbers—by using a "wild thought," as he coin it. While imaginary numbers are effective in mathematics, their use in physics—especially in quantum mechanics and string theory—leads to the paradoxical conclusion that elementary particles, and all matter—including out bodies—do not belong to this physical universe but exist in imaginary, or "quantum," spaces that are explained only mathematically.

As Sir Roger Penrose noted:

"There is an important gap in our physical picture of the world, at a level which ought to bridge

the submicroscopic world of quantum physics to the macro-world of classical physics."

Bombelli's "wild thought" is best understood through Minkowski's 1908 assertion:

"... we would then have in the world no more the space, *but an infinite number of spaces analogously as there is an infinite number of planes in three-dimensional space."*

In Earth sciences, geologist Samuel Warren Carey's Expanding Earth Model is dismissed by the scientific establishment due to *the lack of a mechanism to explain the expansion.* This book explains why mainstream geologists are fundamentally mistaken.

Carey was explicit about this issue:

"The problem is, what triggers the [expansion] process. As a geologist, I insist that the Earth has expanded, and leave it as a cosmological problem of the whole universe. Hence I do not see it as a problem specifically for the Earth, or for the solar system, but for the Universe and Cosmos."

To advance their work and further develop fundamental science in these areas, the author revisits aspects of mathematics (imaginary numbers and quantum spaces), physics (multispace structure of reality), cosmology (multispace reality), and geology (explaining Earth's past, present, future, and expansion).

* * *

CONTENTS

LIST OF FIGURES

Part I

ULTIMATE SECRET OF REALITY

1 ULTIMATE SECRET OF REALITY

We may never discover how the Universe ticks.
Our brains are limited. It may take a posthuman species to work out the big questions.
A true fundamental theory of the universe may exist but could be just too hard for human brains to grasp.
— Professor Martin Rees[35]
British cosmologist and astrophysicist

1.1 WHAT IF?

Many visionaries propose a larger multiverse, but without evidence, these ideas remain speculative. However, what if their visions are accurate, and we have misunderstood reality by viewing it solely as the space of the universe we can see?

What if the reality we perceive is merely a fraction of the true reality, limited by the cosmic horizon of our universe?

True innovation in physics demands a departure from established paradigms, especially from the single space-time framework of Einsteinian relativity. By adopting Minkowski's vision of infinite spaces—analogous to planes within a three-dimensional structure—we shift our understanding toward the ultimate secret of reality: the fundamental ontology of infinite spaces.

In 1572, over three centuries before Minkowski, the mathematician Rafael Bombelli discovered multiple spaces. He introduced $\sqrt{-1}$ (the square root of -1), forming the basis of *imaginary and complex numbers*, which underpin quantum theory. To avoid appearing foolish, he described it as merely a *wild thought*. More about him is in section 8.2 on page 69.

This novel approach treats each space as a distinct entity, with interactions governed by novel principles, without integrating with single-spacetime constraints.

Physicalism, the current paradigm of a unique, single space universe, asserts that reality consists solely of physical matter and energy. However, what we call "physical matter" is not "physical" at all. Quantum physics describes elementary particles, the fundamental components of matter, through mathematical formalisms like wave functions and probability amplitudes, instead of as classical, localized physical objects with definite paths.

In essence, elementary particles—and, in fact, all matter, including our bodies—do not physically exist in this universe. So where do our bodies truly reside?

This is just one example of the many fundamental questions about this reality that remain unanswered by current science.

Inspired by Hermann Minkowski's work, the book begins by revealing the ultimate secrets of reality. The new concept of reality with infinite spaces is validated by uncovering the ultimate secrets of four fundamental pillars: mathematics, physics, cosmology, and geology.

The new paradigm introduces its own terminology.

1.2 MULTISPACE REALITY

The term for a reality with infinite spaces is "Multispace Reality." Using "Multispace" as a singular noun suggests an all-encompassing, limitless realm composed of infinite spaces.

The term "Multispace" was introduced by Florentin Smarandache in 1969 and refers to a framework in which multiple spaces, each potentially governed by its own structure, logic, or physical laws, coexist and interact to describe reality in a unified way [103]. In this model, a multispace is a general mathematical and philosophical environment capable of combining different types of spaces—geometric, physical, logical, or conceptual—each with its own multistructure.

The term aims to transcend disciplinary boundaries by integrating diverse systems, ranging from micro to macro levels in physics and from logic to philosophy, under a single formalism.

Smarandache envisions the multispace as a "Theory of Everything" applicable across all domains of knowledge, connecting gravitational, electromagnetic, weak, and strong interactions in physics, but also extending to cognitive and logical systems through neutrosophic transdisciplinarity, which studies the common properties of apparently unrelated or contradictory entities (such as A, non-A, and neutral-A).

1.3 OMNISPACE

The term *omnispace* describes the *multispace reality as a whole* and may appear speculative due to its absence from mainstream cosmological research. While referring to *reality*—the vast expanse of existence—scientists currently use terms like *multiverse* and *omniverse*, which typically encompass all universes, framing reality within the context of *verses*.

In contrast, envisioning reality as a container or spatial framework, an *omnispace* offers a more comprehensive perspective. This concept serves as the ultimate container for universes and any conceivable or inconceivable structures of existence. By transcending the limitations of current language and conceptual frameworks, which often overlook fundamental forms of existence through a narrow focus on *verses*, *omnispace* challenges restrictive categorical thinking. It broadens our understanding to encompass myriad forms of existence beyond traditional definitions.

We envision the omnispace as a dynamic, multilayered structure, similar to a set of Matryoshka dolls, a collection of wooden figures of decreasing size, each fitting perfectly inside the next larger one. The largest, outermost doll traditionally depicts a Russian woman and separates at the middle to reveal a smaller, similar doll inside. This nesting process repeats for several layers until reaching the smallest solid doll at the center [124].

The omnispace resembles the largest, outermost doll. However, unlike the dolls, each layer contains not just one doll but numerous interconnected spaces, either within the same or different layers. The smallest solid doll at the center represents our universe.

The question is, we do not yet know which layer of the omnispace our universe occupies in the chain of universes.

This nested configuration implies that our universe, other universes, and even elementary particles—which are also independent spaces—are equivalent spacetime frameworks governed by the same physical laws.

1.4 SOLI REALITY

The term *soli reality* refers to the entire existential scope of our single universe, including its unique physical laws and observable phenomena. This term distinguishes our immediate reality from the broader true reality of the omnispace. The prefix *soli-* derives from the Latin *solus,* meaning *alone* or *single,* aptly conveying the idea of a singular universe's reality as opposed to the collective reality of the omnispace.

* * *

Part II

ORIGIN OF MULTISPACE FRAMEWORK

2 ON MINKOWSKI

While there exists an unanimous consensus on the mathematical significance of spacetime for theoretical physics, for a hundred years there has been no consensus on the nature of spacetime itself.
—Vesselin Petkov[89]

Hermann Minkowski (1864–1909), a notable German mathematician and physicist, significantly contributed to the mathematical foundations of special relativity. He taught at various German universities, including Bonn, Königsberg, Göttingen, and the Eidgenössische Polytechnikum (now ETH Zürich). At the Zürich Polytechnic, he was Einstein's mathematics professor in courses such as *Geometry of Numbers*, *Function Theory*, and *Analytical Mechanics* [106].

At the beginning of his career, Minkowski was interested in pure mathematics, investigating quadratic forms and continued fractions. One of his important achievements was the development of *Geometrie der Zahlen* (*Geometry of Numbers*), which was first presented in 1896 and published post-mortem in 1910 [47].

Hermann Minkowski revolutionized our understanding of the universe with his geometric interpretation of special relativity. His insight that space and time form an inseparable four-dimensional continuum reshaped physics and provided the mathematical framework for Einstein's theory of general relativity.

When Einstein published his special theory of relativity in 1905, he presented it primarily as mathematical relationships governing space and time.

However, Minkowski recognized these relationships revealed a deeper truth about reality itself. In his groundbreaking 1908 lecture "Space and Time," he declared that *space and time as independent entities would fade into mere shadows, with only their union maintaining physical reality.* This insight marked a watershed moment in theoretical physics, introducing what we now call "Minkowski spacetime."

The brilliance of Minkowski's contribution lay in recognizing that Einstein's equations could be understood geometrically in a four-dimensional space, with time as the fourth dimension. His mathematical framework treated space and time coordinates as a single four-dimensional continuum, now known as "Minkowski space." Through this geometric interpretation, complex aspects of special relativity became more intuitive and mathematically tractable.

Minkowski is renowned for introducing a two-dimensional spacetime diagram to explain the Lorentz transformation of Special Relativity. This intuitive visual aid has helped clarify Einstein's theory over the last century, significantly facilitating its dissemination to scientists and the general public. His contribution to the development of modern mathematics is enormous, though still not fully understood.

The Lorentz transformation (LT) describes the relationship between two coordinate systems in inertial (i.e., uniform) motion, in terms of their relative speed [116]. The point of origin of each system is a hypothetical observer O, who can make use of physical devices to measure the spacetime coordinates of other events and objects. It is for this reason that coordinate systems in special relativity are also called *reference frames*.

When H. Minkowski began thinking about the structure of the Lorentz group in 1907, he noticed that the geometrical relations between inertial reference frames were not Euclidean (as in classical kinematics), but hyperbolic [119]. However, the spacetime formalism he developed for special relativity did not fully exploit this insight.

According to Scott Walter [118]:

> *"...Minkowski retained the geometric interpretation of the Lorentz transformations that had accompanied the now-banished non-Euclidean interpretation of velocity vectors. In doing so, he elaborated the notion of velocity as a rotation in four-dimensional space. He introduced a formula for the frame velocity... in terms of the tangent of an imaginary angle."*

We reproduce his formula below in modern notation:

$$\beta = \frac{v}{c} = -\mathrm{i}\tan \mathrm{i}\psi = \frac{(e^{\psi} - e^{-\psi})}{(e^{\psi} + e^{-\psi})} \tag{2.1}$$

As Walter points out, Minkowski could have expressed this formula as $\beta = \tanh\psi$, using a real hyperbolic angle instead of an imaginary circular angle. Instead, he chose to express the Lorentz transformation using imaginary trigonometric functions:

$$\begin{aligned} x_1' = x_1, \qquad x_2' = x_2, \\ x_3' = x_3 \cos i\psi + x_4 \sin i\psi, \\ x_4' = -x_3 \sin i\psi + x_4 \cos i\psi \end{aligned} \tag{2.2}$$

Walter elaborates on this choice:

> *"The use of circular functions here underscores the fact that a special Lorentz transformation is equivalent to a rotation in the (x_3x_4)-plane. Likewise, by expressing velocity in terms of an imaginary rotation, Minkowski may have betrayed his knowledge of the formal connection between the composition of Lorentz transformations and relative velocity addition, remarked earlier by Einstein on different grounds (Einstein, 1905, §5). However, Minkowski neither mentioned the law of velocity addition, nor expressed it in formal terms"* [118].

Commenting further he explained:

> *"Minkowski's preference for circular functions may be understood in relation to his project to express the laws of physics in four-dimensional terms. ... Expressing the Lorentz transformation as a hyperbolic rotation would have obscured the connection for physicists"* [118].

Minkowski's geometric approach introduced the concept of world lines—the paths objects trace through four-dimensional space-time. This representation provided physicists with a powerful tool for understanding causality and the relationships between events in space-time. His framework also gave rise to the concept of light cones, which define the causal structure of space-time by showing all possible paths that light signals could take from any given event.

The impact of Minkowski's reformulation of special relativity extends far beyond its initial scope. His geometric framework proved essential for Einstein's development of general relativity, providing the mathematical foundation necessary to understand gravity as the curvature of space-time. Without Minkowski's geometric interpretation, the mathematical formulation of general relativity might have remained out of reach.

Minkowski's influence extends deeply into quantum field theory, where particle physics meets special relativity. Modern quantum field theory is formulated in Minkowski space, and his mathematical structure remains central to our understanding of particle interactions and quantum mechanics. His framework provides the essential tools for calculating particle interactions at high energies, analyzing electromagnetic wave propagation, and describing the motion of objects at relativistic speeds.

The theoretical legacy of Minkowski's work continues to shape contemporary physics research. His insights about the nature of space-time guide modern attempts to reconcile quantum mechanics with gravity. String theory and other unified theories build upon his approach, extending it to include additional dimensions and more complex geometric structures. However, the fundamental insight of treating space-time as a unified geometric entity remains at their core.

In cosmology, Minkowski's framework provides the mathematical foundation for understanding cosmic evolution and the behavior of space-time at the largest scales.

Contemporary cosmological models build upon his geometric approach to describe the large-scale structure of the universe. The mathematical tools he developed remain essential for modeling everything from quantum interactions to cosmic expansion.

The profound significance of Minkowski's contribution lies not only in its mathematical elegance but in how it fundamentally changed our conception of reality. Before Minkowski, space and time were viewed as separate entities, merely providing the stage upon which physical phenomena played out. After Minkowski, they became active participants in physics, inextricably woven together into the fabric of reality itself.

Minkowski demonstrated that the seemingly abstract mathematical relationships in Einstein's special relativity reflected a deeper geometric reality of our universe. His work exemplifies how mathematical insight can lead to profound physical understanding, revealing fundamental truths about the nature of reality. The space-time concept he introduced remains essential to our comprehension of the universe across all scales, from quantum phenomena to cosmological structures.

Modern theoretical physics continues to operate within the conceptual framework Minkowski established. His geometric interpretation of special relativity not only provided the mathematical tools needed for advanced physical theories but also fundamentally changed how we think about the nature of space, time, and the universe itself. This conceptual revolution stands as one of the most significant and enduring contributions to theoretical physics, shaping our understanding of the physical world for over a century.

It is less known that Minkowski had a hidden insight into a deeper physical meaning of his work that he decided not to uncover at the time. For this reason, we also present important aspects of his cryptic legacy.

* * *

3 CRYPTIC LEGACY

Minkowski protested against the use of the word relativity *to describe a theory based on an* absolute *(spacetime), and, had he lived to see the general theory of relativity, I believe he would have repeated his protest in even stronger terms.*
—J. L. Synge [110]

Minkowski presented his groundbreaking theory in the lecture *Raum und Zeit* (Space and Time) at the 80th Assembly of German Natural Scientists and Physicians in Cologne on September 21, 1908.

His famous declaration, *"Henceforth space by itself, and time by itself, are doomed to fade away into mere shadows, and only a kind of union of the two will preserve an independent reality,"* captured the profound paradigm shift in physics.

The theory introduced several key concepts, such as *proper time*, the time experienced by an object in its own frame of reference, and *world lines*, representing an object's path through spacetime. Minkowski also defined the spacetime interval, which remains invariant under Lorentz transformations.

While the lecture was criticized for not being very clear due to its highly mathematical and abstract nature—even Einstein initially found the mathematical approach overly complicated—there may have been purpose in this opacity.

In January 1909, Minkowski tragically passed away from a ruptured appendix at 44, leaving some of his key theoretical insights unexplored. His work provided the mathematical framework for special relativity and laid essential groundwork for Einstein's later development of general relativity. However, some scholars argue that aspects of his original theory remain to be fully understood and utilized.

The *Theory of the Absolute World*, introduced by Minkowski in his lecture, diverges significantly from Einstein's relativity. It is an independent theory where reality comprises a multispace manifold with infinite nested spacetimes and universes. Minkowski described this shift as *truly radical*, but it is more than that. This theory transitions human understanding from the single-space paradigm to a multispace reality, rendering Einstein's relativity outdated.

At the time of this lecture, the theory was still incomplete. As Minkowski explained, *he had independently reached the same conclusions as Einstein but chose not to publish them, as he wanted to first develop the mathematical structure in all its splendor [122].*

This affirmation is supported by Max Born, who attended the seminar on electron theory co-organized by Minkowski and Hilbert in the summer semester of 1905. Many years later (in 1959), Born wrote:

> *"I remember that Minkowski occasionally alluded to the fact that he was engaged with the*

> *Lorentz transformations, and that he was on the track of new interrelationships.*[1] [29]"

The author investigated the time Minkowski spent developing his theory before 1905 by searching the *Jewish National and University Library*, where Minkowski's and Einstein's manuscripts are archived. In Box 9, folder 7 (1904), four pages were found with Minkowski's sketches of orthogonal spaces [79]. (See Figure 3.1 on page 21.)

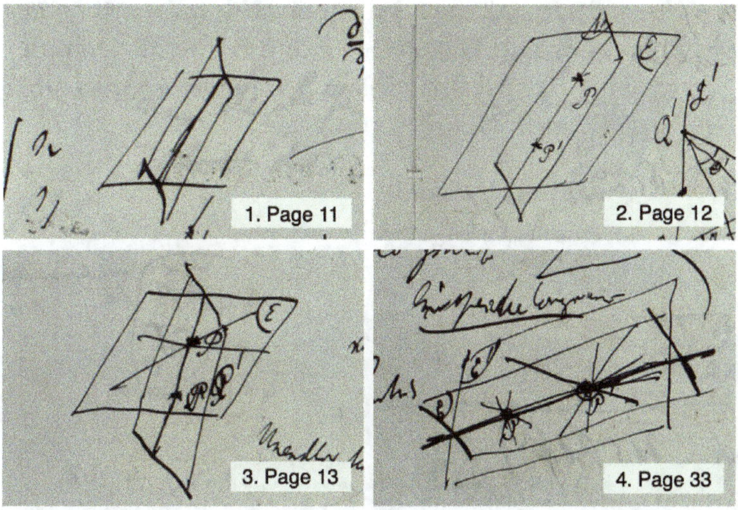

Figure 3.1: Minkowski's four sketches of orthogonal spaces. Credit: Jewish National and University Library.

Based on these sketches, it is plausible that Minkowski developed his theory before 1904. He could not have conceived the orthogonal planes P and P' without prior thought, as evidenced by these sketches.

1 Ich erinnere mich, daß Minkowski gelegentlich Andeutungen machte, daß er sich mit den Lorentz-Transformationen beschäftigte und neuen Zusammenhängen auf der Spur sei.

To conceptualize such a three-dimensional combination, he likely discovered his 2D diagram, introduced in the Cologne lecture in 1908, before understanding the 3D multispace concept. In this lecture, he chose to reveal only the simpler 2D spacetime diagram.

These sketches illustrate the challenges that prevented him from solving, as he described, *the mathematical structure in all its splendor.*

Crucially, to link the 2D diagram with these orthogonal spaces, he would have needed to discover the complex form of the *3D Multispace Diagram*, designed by the author in 2012. Refer to Figure 3.2 on page 22.

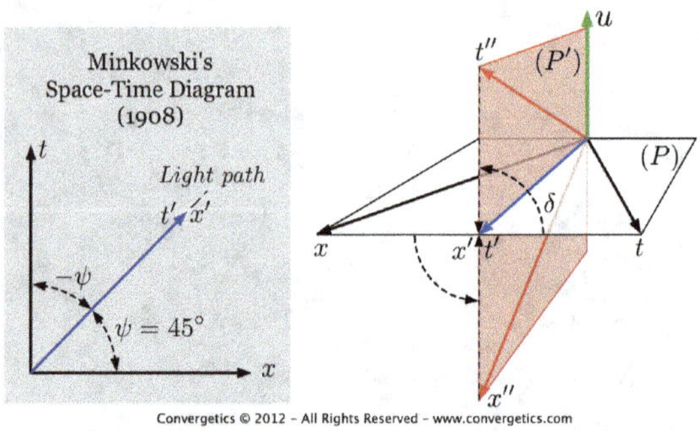

Figure 3.2: The 3D Multispace Diagram.
Credit: CSC (2012) CC BY-NC-ND 4.0.(2012)

It is now clear why Minkowski could not fully express the radical paradigm shift.

To secure his scientific priority, he introduced the theory using cryptic terms like *worldpoint (Weltpunkt), worldline (Weltlinie),* and *world (Welt).* This allowed him to obscure their true significance in the multispace setup by providing accurate yet insufficient definitions.

His clever encryption achieved remarkable results. Physicists and mathematicians worldwide, unaware of the *radical* shift, believed they understood his theory and conducted research supporting relativity based on their incorrect perceptions. Lacking a comprehensive understanding, they deceived themselves for over a century, rendering many scientific efforts futile.

Minkowski likely likely intended to resolve later this mathematical issue and publish the second part of his theory, securing both parts with the priority date of the Cologne lecture, in case someone else developed a three-dimensional diagram from his 2D spacetime diagram. Unfortunately, he died suddenly just four months after this presentation, a loss for the entire scientific community.

* * *

4 | COLOGNE LECTURE

Science is not concerned only with things that we understand. The most exciting and creative parts of science are concerned with things that we are still struggling to understand.
— Freeman Dyson[45]

When Minkowski presented his Cologne lecture in September 1908, he ensured the scientific priority of his theory while concealing enough details to prevent other scientists from fully understanding his discovery [71].

Let us now examine how Minkowski crafted his cryptic theory.[1]

Minkowski began his lecture with a bold declaration, making it clear from the outset that he was going to introduce a revolutionary theory:

"Gentlemen! The views on space and time which I wish to lay before you have sprung from the soil

1 The saying goes: "Traduttore, traditore" ("Translator, traitor"). Following his passing, numerous English translations of his lecture emerged, each imbued with a unique essence based on the translator's interpretation of Minkowski's ideas. Consequently, translating his work posed a formidable challenge for all. Faced with this dilemma, the author opted for a 2009 bilingual edition in German and English of the lecture to verify the accuracy of the translation. In certain instances, he presents both versions for clarity.

> *of experimental physics. Therein lies their strength. Their tendency is radical.*[2]"

Minkowski stated *Their tendency is radical,* but intentionally avoided elaborating that his lecture was *extremely* radical. He was aware that the theory he was about to unveil would introduce a new paradigm of reality.

In reality, despite his claim that the new view was based on experimental physics, Minkowski's theory was rather a very mathematical approach to relativity and, as expected, the discrepancy between the lecture's outset and its content didn't pass unnoticed.

As Scot Walter points out [120]:

> *"The evocation of experimental physics was significant in the first sentence of Minkowski's lecture, and it was deceptive. In what followed, Minkowski would refer to experimental physics only once ..."*

> *"Otherwise, Minkowski kept his promise of a* rein mathematische *exposé, devoid of experimental considerations."*

Indeed, since Minkowski's theory on relativity and the spacetime diagram are both *purely mathematical,* he could not consider them as derived from *experimental physics.* Quite the opposite! However, a three-dimensional spacetime diagram, even if based on 2D geometry, lacked the mathematical foundation. Thus, the only way to introduce it was to assert, as he did, that it *arose from the soil of experimental physics.*

2 M. H.! Die Anschauungen über Raum und Zeit, die ich Ihnen entwickeln möchte, sind auf experimentell-physikalischem Boden erwachsen. Darin liegt ihre Stärke. Ihre Tendenz ist eine radikale.

In fact, Minkowski proposed a fundamental change in our understanding of space and time:

> *"Henceforth space by itself, and time by itself, are doomed to fade away into mere shadows, and only a kind of union of the two will preserve an independent reality on its own."*

This statement reflects Minkowski's belief that his reformulation of space and time into a *spacetime manifold* was not merely mathematical but deeply rooted in physical observations, and importantly, *completely different* from the unique spacetime continuum paradigm of reality.

As Walter notes, until now, this statement was seen as merely a *grandiose announcement* of his theory [120] in [52].

However, when we read Minkowski's theory, we see that uniting space and time into spacetime does not make either space or time fade into *mere shadows*.

He continues by explaining:

> *"First of all, I would like to show how it might be possible, setting out from the adopted mechanics of the present day, along a purely mathematical line of thought, to arrive at changed (i.e. new) ideas of space and time."*

Further on, Minkowski details his thinking by illustrating these relations using a mathematical graphic method:

> *"Let* x, y, z *be orthogonal coordinates for space, and let* t *denote time. The objects of our perception invariably include places and times in combination. Nobody has ever noticed a place except at a time, or a time except at a place. But I respect*

> *the dogma that space and time each have an independent meaning. I will call a point in space at a given time, i.e., a system of values* x, y, z, t, a worldpoint (ein Weltpunkt)."

We observe that Minkowski introduced the spacetime concept by naming this manifold a *worldpoint (ein Weltpunkt)* to designate it as the smallest spacetime building block of reality.

This simple concept is essential to understanding all of Minkowski's multispace theory. The *Weltpunkt* is like an atom for matter or a brick used to build a house. Minkowski constructs his entire theory beginning with it, and it is crucial for anyone wanting to comprehend the existence of reality.

He generalizes it to encompass all of reality:

> "*The* manifold *of all possible systems of values* x, y, z, t *will be called* the world (die Welt)."

This explanation is crucial because he defined *his world* to represent the entirety of reality.

Notably, Minkowski did not define the *world* as a *spacetime continuum,* as Einstein later did, but as a *manifold*[80], a whole consisting of diverse elements, meaning a *world* composed of *an infinite number of* Weltpunkte, *(i.e., spaces),* as he stated at the end of the first part of his lecture.

> "*... we would then have in the world no more* the space, *but an infinite number of spaces analogously as there is an infinite number of planes in three-dimensional space. Three-dimensional geometry becomes a chapter in four-dimensional physics.*"

He ended this section with a comment:

> *"I think the word relativity postulate used for the requirement of invariance under the group Gc is very feeble. Since the meaning of the postulate is that through the phenomena only the four-dimensional world in space and time is given, but the projection in space and in time can still be made with certain freedom, I want to give this affirmation rather the name "the postulate of the absolute world" (or shortly the world postulate)."*

Unfortunately, Minkowski died soon after that lecture without clearly explaining what he meant by *an infinite number of spaces*. Not recognizing the significance of Minkowski's declaration, no scientist—until now—took his hint seriously, dismissing it as a *grandiose announcement* of the mathematician's own relativity theory.

Einstein initially viewed Minkowski's treatment as a mere mathematical trick, before eventually realizing that a geometrical view of spacetime was necessary to complete his later work in general relativity (1915) [118].

The statement about the *infinite number of spaces* in our universe is crucial for understanding Minkowski's theory. His seemingly counterintuitive logic challenges common sense, which explains why it puzzled Einstein and his peers. This is where Minkowski's brilliance shines.

In Minkowski's postulate, *worldpoints* are the fundamental elements of the world. He explains:

> *"Not to leave a yawning void anywhere, we will imagine that everywhere and at all times there is something perceptible. To avoid saying matter or electricity, I will use for this something the word*

> substance (die Substanz). *We focus our attention on the substantial point existing at the worldpoint* x, y, z, t *and imagine that we can recognize this substantial point at any other time."*

Minkowski clarifies that he refers not to mathematical abstractions of space and time, but to actual physical spacetimes, which he calls *substantial points*, encompassing their entire histories.

Minkowski further expands this point-like manifold into a real spacetime:

> *"Corresponding to a time element* dt, *we have the variations* dx; dy; dz *of the space coordinates of this substantial point. Then we obtain, as an image, so to speak, of the eternal course of the substantial point, a curve in the world,* a worldline (eine Weltlinie), *whose points can be clearly related to the parameter t from* $-\infty$ *to* $+\infty$*."*

To eliminate any doubts about the physicality of these substantial spaces, Minkowski described *worldlines* as collections of all worldpoints associated with a substantial point when *t* takes all possible values.

Minkowski establishes that each *worldline* is an independent and eternal spacetime manifold within the world manifold.[3]

As noted by Vesselin Petkov [88]:

3 In current physics, a worldline is a path traced by an object through 4-dimensional spacetime, representing the history of its location in space over time. Physicists coined this term because they could not fully grasp Minkowski's original meaning.

> *"Minkowski... believed that spacetime is not merely an abstract four-dimensional mathematical space, but represents a real four-dimensional world of one temporal and three spatial dimensions. In this* world, *the perceived macroscopic three-dimensional bodies are real four-dimensional* worldlines, *or rather* worldtubes *since we are dealing with spatially extended physical bodies."*

To highlight the importance of these *worldlines* in his theory, Minkowski used this definition to announce the groundbreaking discovery that reality is a multispace *manifold*[80]. He did not describe reality as a unique *spacetime continuum*, as Einstein later did.

In Minkowski's view, this "world manifold" consists of countless spacetime manifolds:

> "*The whole world presents itself as resolved into such* worldlines, *i.e., physical spacetimes, and I want to say in advance, that in my understanding the laws of physics can find their most complete expression as interrelations between these* worldlines."

Minkowski's groundbreaking declaration becomes clearer when expressed in contemporary language:

> "*The entire reality appears to resolve itself into such independent spacetimes. I would like to say that, in my opinion, the laws of physics find their most perfect expression in the interrelationships between these independent spacetimes.*"

Let us highlight the difference: *Reality is based on interrelationships between independent spacetimes, not between objects in a unique spacetime, as we now think.*

Finally, to ensure he unequivocally conveys the correct message, even if no one understands his unfinished theory, Minkowski concludes the first part of the lecture with a statement that is cryptic in Einstein's paradigm but perfectly clear in Multispace:

> *"We would then have in the world no longer* the space, *but an infinite number of spaces, analogously as there are in three-dimensional space an infinite number of planes. Three-dimensional geometry becomes a chapter in four-dimensional physics. You see why I said at the outset that space and time are to fade away into shadows, and that only a world in itself will subsist."*[4]

Not understanding the extreme importance of Minkowski's declaration, most physicists underestimated its significance, dismissing it as merely a grandiose claim of his own relativity theory.

However, Minkowski's final statement in the first part of "Raum und Zeit" lecture strongly supports the Author's interpretation of his theory.

Minkowski was the pioneer of the multispace reality.

4 Hiernach würden wir dann in der Welt nicht mehr den Raum, sondern unendlich viele Räume haben, analog wie es im dreidimensionalen Räume unendlich viele Ebenen gibt. Die dreidimensionale Geometrie wird ein Kapitel der vierdimensionalen Physik. Sie erkennen, weshalb ich am Eingange sagte, Raum und Zeit sollen zu Schatten herabsinken und nur eine Welt an sich bestehen.

This paradigm shift is monumental for theoretical physics and science as a whole.

Regrettably, Einstein overlooked this conclusion when developing general relativity, causing a century-long drift in physics.

On a positive note, Einstein resisted the probabilistic interpretation of quantum mechanics. In 1926, he asserted that quantum mechanics, despite its mathematical success, couldn't fully describe physical reality:

> *"I, at any rate, am convinced that He does not play dice."* [16][5]

Einstein's skepticism was rooted in the philosophical implications regarding reality's nature. His 1935 *EPR paper* questioned quantum mechanics' completeness, suggesting the theory lacked elements necessary for a full description of physical reality [38].

He was correct. The Multispace paradigm of reality clarifies quantum physics through these independent spaces.

* * *

5 Jedenfalls bin ich überzeugt, daß der Alte nicht würfelt.

5 SCIENTISTS ON MINKOWSKI'S THEORY

Minkowski protested against the use of the word relativity *to describe a theory based on an* absolute *(spacetime), and, had he lived to see the general theory of relativity, I believe he would have repeated his protest in even stronger terms.*

—J. L. Synge [110]

5.1 ALBERT EINSTEIN

In contrast to Minkowski's *absolute world filled with an infinite number of spaces*, Einstein's 1916 postulate in *Relativity: The Special and General Theory* states [40]:

> "*. . . the world of physical phenomena, briefly called* world *by Minkowski, is naturally four-dimensional in the spacetime sense. It is composed of individual events, each described by four numbers: three space coordinates* x, y, z, *and a time coordinate, the time value* t.
>
> *The* world *is also a continuum in this sense; for every event, there are as many* neighboring *events (realized or at least thinkable) as we choose, with coordinates* x1, y1, z1, t1 *differing by an indefinitely small amount from those of the event* x, y, z, t *originally considered.*"

While acknowledging that the world *is composed of individual events,* Einstein quickly concluded that the *"world" is in this sense also a continuum.* In Chapter 24, *Euclidean and non-Euclidean Continuum,* he further explains his logic:

> *"The surface of a marble table is spread out in front of me. I can get from any one point on this table to any other by passing continuously from one point to a* neighboring *one, repeating this process a (large) number of times, or, in other words, by going from point to point without executing* jumps. *I am sure the reader will appreciate with sufficient clearness what I mean here by* neighboring *and by* jumps *(if he is not too pedantic). We express this property of the surface by describing it as a continuum."*

We do not intend to diminish Einstein's genius or work. However, by concentrating only on Minkowski's *worldpoints,* he did not progress beyond the fundamentals of Minkowski's 4D concept. By overlooking the *worldlines,* the essence of Minkowski's discovery, Einstein erroneously assumed the universe is a singular spacetime continuum.

Consequently, while developing general relativity, he overlooked Minkowski's insight that *the laws of physics* (including gravity!) *are best expressed as interrelations between these worldlines,* i.e. real spacetimes.

Had Einstein fully understood Minkowski's theory, he might not have developed general relativity as he did, nor spent his life pursuing the unification of gravity with the other three fundamental forces.

5.2 VESSELIN PETKOV

Reflecting on the impact of Minkowski's premature death, Vesselin Petkov speculates:

> *"We will never know how physics would have developed had he lived longer. What seems undeniable is that the discovery of the true cause of gravitation*—the non-Euclidean geometry of spacetime—*would have been different from what actually happened."*

Petkov hypothesizes about Minkowski's perspective on gravitation:

> *"...As a mathematician, he would be appalled by what he saw as a confusion of physics and geometry:*
> • *The new theory of gravitation shows that gravitational physics is actually the geometry of curved spacetime; no general relativity of anything can be found there;*
> • *How could physicists claim that within general relativity, gravitational phenomena are caused by gravitational interaction? According to general relativity, gravity is not a physical interaction...*
> • *How could physicists discuss gravitational energy within general relativity? There is no gravitational field or force; the gravitational field is at best geometric, not physical, and thus possesses no energy."*

It is important to note that Petkov's assumptions are not baseless. His view is supported by J. L. Synge, who remarked [110]:

> *"Minkowski protested against using the word* relativity *to describe a theory based on an* absolute *(spacetime), and, had he lived to see the general theory of relativity, I believe he would have repeated his protest even more strongly."*

We must also acknowledge two facts that may explain why Minkowski's *Postulate of the absolute world* remains largely misunderstood by physicists, including Einstein, to this day.

The theory is surprisingly complex and distinct from special relativity, which explains why Einstein and many other physicists struggled to grasp its full complexity. The examples mentioned are just a few instances where other scientists, who understood Einstein's mistake but were less prominent, could not effectively communicate his error.

5.3 ARTHUR STANLEY EDDINGTON

Arthur Stanley Eddington was a British astronomer, physicist, and mathematician who was an early advocate of Einstein's General Relativity. He is renowned for the expedition to observe the 1919 Solar eclipse, which provided one of the earliest confirmations of relativity.

His example is relevant, as he made a clear comment on the reality of spacetime when discussing the fact (revealed by Minkowski) that *not only do observers in relative motion have different times, but they also have different spaces,* while emphasizing *which however are fictitious since according to the theory of relativity the world is not objectively divided into such spaces and times* [37]:

> *"It was shown by Minkowski that all these fictitious spaces and times can be united in a single continuum of four dimensions. The question is often raised whether this four-dimensional spacetime is real, or merely a mathematical construction; perhaps it is sufficient to reply that it can at any rate not be less real than the fictitious space and time which it supplants."*

In essence, Eddington's point is that if Einstein had not imposed the idea that *the world is a unique spacetime continuum,* physicists might have eventually concluded that these *different spaces* were not so *fictitious* after all.

As Eddington said: *perhaps it is sufficient to reply that it can at any rate not be less real than the fictitious space and time which it supplants,* i.e. the spacetime continuum.

While we are ready to end this chapter, one may wonder what is the point of discussing Minkowski.

We know the fundamental incompatibility between general relativity and quantum mechanics, the two cornerstone theories of modern physics. These theories describe the universe at vastly different scales—general relativity governs the macroscopic realm of gravity and spacetime, while quantum mechanics governs the microscopic realm of particles and their interactions.

Scientists still do not know the reasons for their incompatibility, and which theory (if either) might be wrong, or if they are both wrong, which is the best possibility.

The core issue arises when attempting to combine these theories into a unified framework, often referred to as *a quantum theory of gravity.*

On the one side, quantum spaces are explained mathematically as imaginary spaces, but are not explained physically. Scientists do not know how to connect them with the physical space of the universe.

On the other side, Petkov envisioned Minkowski's possible take on gravity: *that gravitational physics is, in fact, the geometry of curved spacetime; no general relativity of anything can be found there.*

In this way, we are heading toward the conclusion that quantum spaces are real and that gravity is a chimera. However, it is still a long way, as these hints are far from a consistent theory. So, while keeping these hints in mind, we will examine the development of *Minkowski's unfinished theory* in the next chapter.

* * *

Part III

ULTIMATE SECRETS OF MATHEMATICS

6 MULTISPACE ALGEBRA

6.1 MINKOWSKI'S DIAGRAM

In 1908, Minkowski introduced his Cologne lecture, although his theory was incomplete. As Minkowski explained, he aimed to first develop the mathematical structure *in all its splendor* [122]. Despite extensive searches, we found only the four pages containing Minkowski's sketches of orthogonal spaces in Figure 3.1 on page 21, along with the following drawing:

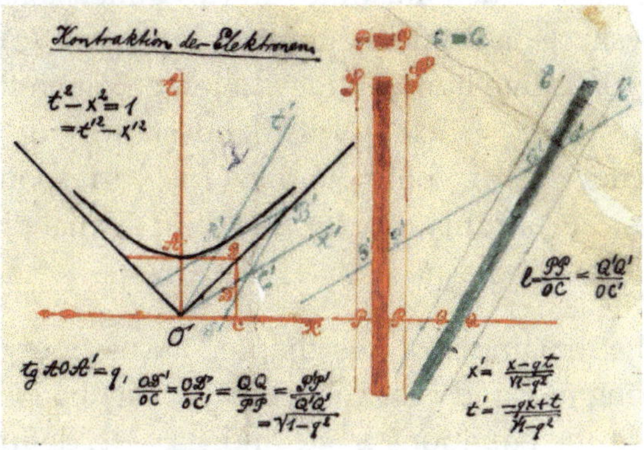

Figure 6.1: Minkowski's spacetime diagram (1908)

Figure 6.2 on page 44 is the modern redraw of this graph.

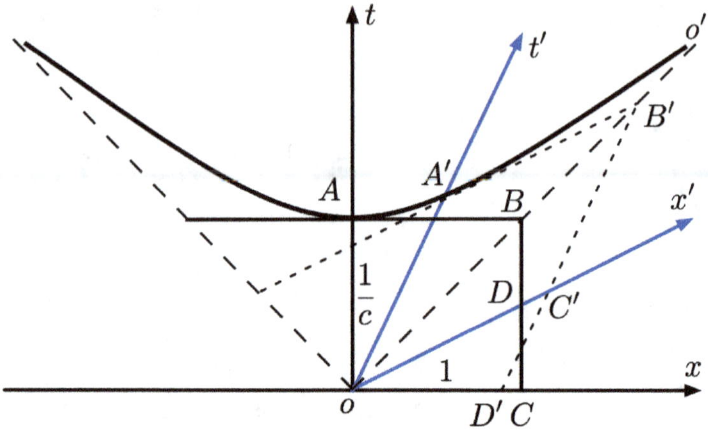

Figure 6.2: Modern spacetime diagram of Minkowski's left side

The hyperbolic nature of the Lorentz transformation becomes evident when illustrated in two dimensions. Using Minkowski's spacetime diagrams (see Figure 6.2), the invariant spacetime interval traces a hyperbola in the (x, t) plane: $(ct)^2 - x^2 = 1$ [118]. This figure shows only the two coordinates x_3 and x_4 of Equation (2.1), corresponding to the physical coordinates x (or any fixed spatial direction) and ct respectively.

For the remainder of this chapter, we assume c equals 1, referring to x and t for convenience. The diagonal lines represent the equations $x = t$ and $x = -t$, defining the paths of observable light rays in empty space. In a fully 4-D spacetime, these light rays form an invariant hypercone [70].

What is the meaning of Minkowski's imaginary angle $\widehat{AoA'}(\mathrm{i}\psi)$? To clarify the difference between a real and an imaginary angle, Figure 6.3(a) shows the circular rotation of a plane $P'(x', t')$ through a real angle ψ relative to plane $P(x, t)$. Figure 6.3(b) depicts a rotation of P' with the imaginary angle $\mathrm{i}\psi$. Figure 6.3(b) simplifies the right-hand side of Figure 6.2.

Let's examine Figure 6.3, starting with the premise that the two coordinate systems are superposed before any rotation occurs. In Figure 6.3(a), the real angle ψ rotates both axes $\mathbf{x}'$ and $\mathbf{t}'$ in the same direction (counter-clockwise for a positive angle). Plane P' clearly remains in the same 2D space as plane P.

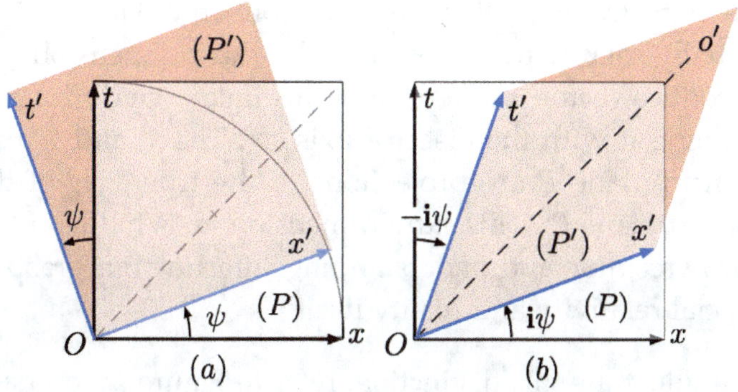

Figure 6.3: The difference between a 2D real rotation (a) and a 2D imaginary rotation (b)

To an observer O within P, as in Figure 6.3(b), the Lorentz transformation rotates the axes $\mathbf{x}'$ and $\mathbf{t}'$ of plane P' in opposite directions. Minkowski, Einstein, and subsequent physicists and mathematicians have regarded this as an imaginary angle. The evidence is clear: the axes $\mathbf{x}'$ and $\mathbf{t}'$ converge towards their bisector oo'.

However, Minkowski's diagram presents a challenge that they all overlooked: the axes $\mathbf{x}'$ and $\mathbf{t}'$ must appear orthogonal to an observer $\mathbb{O}'$ within P'. This orthogonality is not apparent in Minkowski's 2D diagram, where P and P' are coplanar.

The author noted this discrepancy and pursued a logical geometric explanation to enhance understanding of the Lorentz transformation. The imaginary rotation in Figure 6.3(b) fails to meet the orthogonality condition in P'. Thus, plane P' undergoes a more intricate rotation.

When the two axes of frame P' are perceived as coplanar with frame P, they seem to converge. However, P' actually rotates away from P into a third dimension. We introduce vector $\mathbf{u}$ to represent this unknown dimension, aligning it with the bisector axis oo'. The $\mathbf{x}'$ and $\mathbf{t}'$ axes seen in plane P are projections of the true axes of the rotated plane P'. This extra dimension, $\mathbf{u}$, is hidden from our perception and understanding, affecting our grasp of special relativity and reality itself.

To illustrate this distinction, refer to Figure 6.4 on page 47, which clearly demonstrates this disparity.

We often consider rotations in space, but the rotation of P' away from P, while physically real, is not a rotation into a spatial dimension. Instead, it is a relativistic rotation into $\mathbf{u}$, linked to the relative speed between the two frames. This raises the question of whether $\mathbf{u}$ can be considered a dimension by the standard physics definition.

Alternatively, $\mathbf{u}$ can be seen as a new degree of freedom for spacetime frames, not for the objects within them. Thus, we must carefully distinguish between a physical rotation of an object within a frame and a relativistic

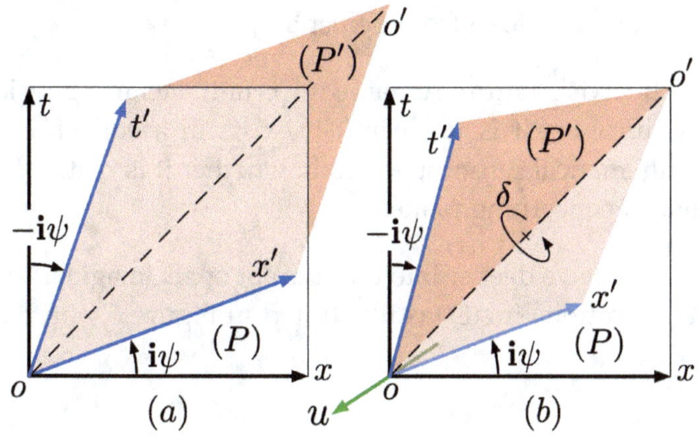

Figure 6.4: The difference between Minkowski's 2D imaginary rotation (a) and a 3D real rotation (b)

rotation into dimension **u** of one spacetime frame relative to another.

Interestingly, the real rotation into the new dimension **u** was not discovered by Minkowski, Einstein, or the many physicists and mathematicians who followed. However, it seems Minkowski understood this concept, though he did not have the chance to fully develop it.

Put simply, the planes P and P' are separate and independent spaces. Minkowski's diagram does not depict a true imaginary rotation in 2D. Instead, it shows a partial view of a real rotation into a new dimension, appearing imaginary when projected onto plane P.

The rotation of plane P' seems imaginary to an observer O in plane P. However, this motion is a rotation with the real angle δ outside of plane P. The spacetime of plane P' physically exits the spacetime of plane P and enters an additional, yet unknown extradimensional space $\mathbf{u}$, previously hidden from perception.

The axis's nature remains uncertain, lacking physical significance. It is unclear if it exists in a conventional mathematical sense, specifically whether it is real or imaginary concerning plane P.[1]

Let's delve deeper into the concept of 2D imaginary-yet-real rotations in 3D, as illustrated in Figure 6.5 on page 49.

6.2 MULTISPACE DIAGRAM

The right side of Figure 6.5 presents another version of Figure 6.4(b). This time, we view the two planes from a vantage point above P.

The author introduced this new representation of the Lorentz transformation as *the Multispace Diagram (MSD)*.

Figure 6.5 shows the horizontal plane P, with the $\mathbf{x}$-axis aligned with segment $\overline{AO}$ and the $\mathbf{t}$-axis aligned with segment $\overline{OB}$. The bisector axis $\overline{OO'}$ divides these segments equally. Plane P' is rotated into the third dimension by real angle δ, positioning its $\mathbf{x}''$-axis along $\overline{A''O}$ and its $\mathbf{t}''$-axis along $\overline{OB''}$. The $\mathbf{x}'$ and $\mathbf{t}'$ axes in P are projections of the $\mathbf{x}''$ and $\mathbf{t}''$ axes, represented by line segments $\overline{A'O}$ and $\overline{OB'}$. It is important to note that the angle ψ between the $\mathbf{x}'$-axis and the $\mathbf{x}$-axis is not the same as the angle δ

1 In mathematics, the term *imaginary* refers strictly to algebraic properties. Thus, the third dimension $\mathbf{u}$ (as shown in Figure 6.5) would only be considered imaginary if the multiplication of its basis element $\mathbf{e}_u$ satisfied the property $\mathbf{e}_u^2 = -1$.

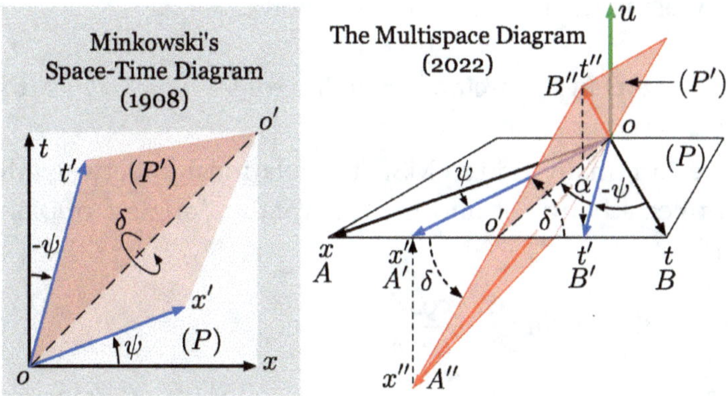

Figure 6.5: The plane P' rotation with respect to plane P.

between P' and P. Our goal is to establish a relationship between these angles.

The angle δ is defined by segments $\widehat{BO'B''}$. The angle ψ is defined by segments $\widehat{B'OB}$. Without loss of generality, we can define segments in P and P' such that:

$$\overline{AO'} = \overline{O'B} = \overline{A''O'} = \overline{O'B''} = \overline{OO'} = 1 \tag{6.1}$$

The right triangle $O'B'B''$ includes the endpoints of the $\mathbf{t}''$- and $\mathbf{t}'$-axes. We can write

$$\overline{O'B'} = \overline{O'B''} \cos\delta = \cos\delta \tag{6.2}$$

Segment $\overline{O'B'}$ also occurs in the triangle $O'OB'$. The value of the angle α in Figure 6.5 is given by the relation

$$\tan\alpha = \frac{\overline{O'B'}}{\overline{OO'}} = \cos\delta \quad \Rightarrow \quad \alpha = \arctan(\cos\delta) \tag{6.3}$$

These angles can also be related to ψ, since

$$\psi = \frac{\pi}{4} - \alpha = \arctan 1 - \arctan(\cos\delta) \tag{6.4}$$

By using the trigonometric relation

$$\arctan x - \arctan y = \arctan \frac{x-y}{1+xy} \tag{6.5}$$

we can finally find in MSD the relationship between the projected angle ψ and the real rotation angle δ of plane P':

$$\psi = \arctan \frac{1-\cos\delta}{1+\cos\delta} \tag{6.6}$$

Equation (6.6) shows a correlation between two distinct real angles, ψ and δ. This aligns with our expectation that the projection of a real angle remains a real angle. However, the tilted plane introduces an imaginary rotation in its projected coordinate system: the $(\mathbf{x}', \mathbf{t}')$ axes have moved closer.

Trigonometric functions with imaginary angles became essential when the Lorentz transformation was initially viewed as a two-dimensional operation. Extending this to three-dimensional space reveals that the Lorentz transformation behaves intuitively.

What can we do with the additional dimension $\mathbf{u}$? One approach is to propose that angles ψ and δ are imaginary, not real. In exploring this, recall Equation (2.1) on page 13, which relates the relative speed v to Minkowski's imaginary angle $\mathbf{i}\psi$. Thus, we can express:

$$\beta = -\mathbf{i} \tan \mathbf{i}\psi \;\Rightarrow \tag{6.7}$$

$$\mathbf{i}\psi = \arctan \mathbf{i}\beta = \mathbf{i} \arctan \frac{1-\cos\delta}{1+\cos\delta} \tag{6.8}$$

using the result of Equation (6.6).

The above formula gives the relationship between the relative speed v and the angle δ belonging to the dimension $\mathbf{u}$.

$$\mathbf{i}\beta = \frac{1 - \cos\delta}{1 + \cos\delta} \quad \Rightarrow \quad \beta = -\mathbf{i}\frac{1 - \cos\delta}{1 + \cos\delta} \tag{6.9}$$

Since β is a real quantity, Equation (6.9) suggests that $\mathbf{u}$ may be imaginary with respect to plane P. If the axis were real, then δ should logically be real as well. Mathematically, the *lost* factor of $\mathbf{i}$ on the right-hand side of Equation (6.9) indicates that the Lorentz transformation does not represent a real rotation in three dimensions.

6.3 MULTISPACE ALGEBRA

We find ourselves in a paradoxical scenario where two seemingly real quantities, β and δ, exhibit an imaginary relationship. However, the investigation of the nature of $\mathbf{u}$ becomes more intricate. While we have previously considered these rotations in relation to the real coordinates $\mathbf{x}$ and $\mathbf{t}$, it is crucial to acknowledge that P and P' represent complex spaces in a deeper physical sense.

This fundamental yet unexplained phenomenon led Einstein and others to adopt the previously obscure Lorentz transformation as a physical law. This interpretation of the Lorentz transformation between inertial reference frames suggests a novel physical coordinate, potentially revealing an unknown aspect of reality. This new coordinate lies outside the plane P and is defined by its basis element $\mathbf{e}_u$, orthogonal to the basis elements $\mathbf{e}_x$ and $\mathbf{e}_t$ of $\mathbf{x}$ and $\mathbf{t}$, respectively.

This concept can be visualized as a three-dimensional space comprising the real plane $P = (\mathbf{x}, \mathbf{t})$ and a third coordinate $\mathbf{u}$. Theoretically, depending on whether $\mathbf{u}$ is real or imaginary, the resulting three-dimensional space can be real or complex. [2]

Continuing our investigation, let us reconsider Equation (6.9) on page 51:

$$i\beta = \frac{1 - \cos\delta}{1 + \cos\delta}. \tag{6.10}$$

When we do the math we obtain[3]

$$\delta = \arccos\frac{1 - i\beta}{1 + i\beta}\ , \text{ where } \beta = \frac{v}{c} \tag{6.11}$$

From the equation above, $\delta = f(v)$ is evident. However, v is real, whereas δ is complex. To comprehend this phenomenon, we make the denominator of the equation real by writing

$$\frac{1 - i\beta}{1 + i\beta} \cdot \frac{1 - i\beta}{1 - i\beta} = \frac{(1 - i\beta)^2}{1 - (i\beta)^2} = \frac{1 - 2i\beta - \beta^2}{1 + \beta^2} \tag{6.12}$$

so that Equation (6.11) becomes:

$$\delta(v) = \arccos\frac{1 - 2i\beta - \beta^2}{1 + \beta^2}, \tag{6.13}$$

---

2 An astute reader may already be aware of the nature of three-dimensional complex spaces. For the purposes of this exercise, we shall disregard our knowledge of Sir William Rowan Hamilton and his groundbreaking discovery of quaternions.

3 The detailed calculations are:

$$i\beta(1 + \cos\delta) = 1 - \cos\delta \Rightarrow i\beta + i\beta\cos\delta = 1 - \cos\delta$$
$$\cos\delta + i\beta\cos\delta) = 1 - i\beta \Rightarrow \cos\delta(1 + i\beta) = 1 - i\beta$$
$$\cos\delta = \frac{1 - i\beta}{1 + i\beta} \Rightarrow \delta = \arccos\frac{1 - i\beta}{1 + i\beta}\ \diamond$$

showing that δ is a complex angle in and of itself.

Separating the arguments of Equation (6.13) into real and imaginary parts, we obtain:

$$\begin{aligned} \delta(v) &= \arccos\left[\frac{1-\beta^2}{1+\beta^2} - \mathbf{i}\frac{2\beta}{1+\beta^2}\right] \\ &= \arccos\left[\Re(v) + \mathbf{i}\cdot\Im(v)\right] , \end{aligned} \tag{6.14}$$

where we have defined

$$\Re(v) = \frac{1-\beta^2}{1+\beta^2} \quad \text{and} \quad \Im(v) = -\frac{2\beta}{1+\beta^2}. \tag{6.15}$$

To see whether Equation (6.14) has any sense, let us validate this result in two critical situations: when $v = 0$, and when $v = c$.

When $v = 0$, $\beta = 0$, and Equation (6.14) becomes:

$$\begin{aligned} \delta(v)_{v=0} &= \arccos\left[\frac{1-\beta^2}{1+\beta^2} - \mathbf{i}\frac{2\beta}{1+\beta^2}\right] \\ &= \arccos\left[\frac{1-0}{1+0} - \mathbf{i}\frac{2\cdot 0}{1+0}\right] \\ &= \arccos\left[\,1 - \mathbf{i}\cdot 0\,\right] \\ &= \ 0^\circ - \mathbf{i}\cdot 90^\circ \end{aligned} \tag{6.16}$$

In the second situation, when $v = c$, $\beta = 1$, calculating Equation (6.14) we obtain:

$$\begin{aligned} \delta(v)_{v=c} &= \arccos\left[\frac{1-\beta^2}{1+\beta^2} - \mathbf{i}\frac{2\beta}{1+\beta^2}\right] \\ &= \arccos\left[\frac{1-1}{1+1} - \mathbf{i}\frac{2\cdot 1}{1+1}\right] \\ &= \arccos\left[\,0 - \mathbf{i}\cdot 1\,\right] \\ &= 90^\circ - \mathbf{i}\cdot 0^\circ \end{aligned} \tag{6.17}$$

To understand the outcomes of these scenarios, it is essential to first grasp the concept of dual space introduced in Chapter 9 *Physical Nature of Imaginaries* on page 85.

This concept suggests the existence of two observers, $\mathcal{O}_1$ and $\mathcal{O}_2$, situated in mutually orthogonal spaces, $\mathcal{S}_1$ and $\mathcal{S}_2$.

Equation 6.16 represents the position of observer $\mathcal{O}_1$ within his space, $\mathcal{S}_1$. Here, the velocity v is zero, and for $\mathcal{O}_1$, this space is real and *here*. For him, space $\mathcal{S}_2$ is imaginary and *there*.

Similarly, Equation 6.17 describes the position of observer $\mathcal{O}_2$ in her space $\mathcal{S}_2$, which has $v = c$ relative to $\mathcal{S}_1$, and is real and *here* for her. Due to its speed, $\mathcal{S}_2$ is rotated into dimension $\mathbf{u}$ by 90° with respect to space $\mathcal{S}_1$ at rest (0°), making it imaginary and *there* for $\mathcal{O}_2$.

Remark 6.3.1. *Considering the results of the two extreme cases ($v = 0$ and $v = c$), we infer that any value of $\{\, v \mid 0 < v < c\}$ results in a complex argument in the function* $\arccos[\, v\,]$. *This indicates that the physical meaning of Equation 6.14 on page 53 is that the Lorentz transformation with imaginary numbers describes the* dual-space *with components* $\mathcal{S}_1$ here *and* $\mathcal{S}_2$ there.

Remark 6.3.2. *When two reference frames, $\mathcal{S}_1$ and $\mathcal{S}_2$, have a relative speed of $+v$, the spacetimes are no longer coplanar, as understood by scientists. The spacetime plane of the moving reference frame $\mathcal{S}_2$ rotates into a third, unseen extradimension $\boldsymbol{u}$ through an angle $\delta = f(v)$ around its light path—the bisector $x = ct$. The real part of the Lorentz transformation of the moving frame represents the projection of this spacetime plane onto the observer's $\mathcal{O}_1$ coordinates. The imaginary part of the* *transformation is perceived from the observer's* O_2 *perspective while in* S_2.

Observers in P cannot directly measure events in P'. Even if they believe they have a direct view of P', they only observe its projection onto their own frame P.

In everyday situations, when $v << c$, observers perceive no difference between directly observing events in P' and observing their projection on P. The distinction between the two frames becomes apparent only at higher velocities: as relative speed increases, the angle of rotation grows, and the projection of P' on P diminishes.

Observers in P' perceive no reduction in measurement values related to events in P', regardless of P''s velocity relative to P.

The relativistic rotation of reference frames, which are independent spacetimes, illustrates Minkowski's insight that *"space and time are not the most fundamental elements of reality."* However, the nature of dimension **u** remains an area of ongoing research and inquiry.

6.4 CORE PRINCIPLE

The Lorentz transformation is a real rotation in a higher-dimensional space involving u, not an imaginary rotation in 2D. The 2D spacetime diagram is a projection, and u represents a hidden dimension tied to relative velocity, potentially complex in nature.

* * *

Part IV

ULTIMATE SECRETS OF PHYSICS

7 ON IMAGINARY NUMBERS

I think I can safely say that nobody understands quantum mechanics. [...] Do not keep saying to yourself, if you can possibly avoid it, 'But how can it be like that?' because you will get 'down the drain,' into a blind alley from which nobody has yet escaped. Nobody knows how it can be like that.
—Richard Feynman

There is an important gap in our physical picture of the world, at a level which ought to bridge the submicroscopic world of quantum physics to the macro-world of classical physics.
—Sir Roger Penrose

If quantum mechanics has not profoundly shocked you, you have not understood it yet. Everything we call real is made of things that cannot be regarded as real.
—Niels Bohr

Elementary particles, the fundamental constituents of all matter in the universe, exist in quantum spaces. They are mathematically described as existing in these spaces, though scientists refrain from labeling them as imaginary. The reality is that no one can physically explain these spaces as real matter in objective reality or determine their location within the universe.

There is a dichotomy and knowledge gap between the scientific fields that describe real matter and those that describe elementary particles. This suggests that humans have a fundamentally flawed perception of the world, hindering a correct interpretation of reality as a whole. Classical mechanics applies to the macroscopic world, which we experience firsthand, even with the aid of microscopes, telescopes, and various other *tools* to enhance our senses. We perceive this physical world as independent, objective, deterministic, and, most importantly, real. Hence, the surrounding world is called reality.

Quantum mechanics studies the structure and behavior of matter's fundamental building blocks and their interactions with energy in the atomic and subatomic realms. The term quantum indicates that in this world, matter particles and energy fields change in discrete amounts (quantum, pl. quanta, in Latin), not continuously. We intuitively expect matter, as a key element of reality, to remain real even when fragmented into its smallest pieces. Here, nature challenges us, as something unexplained occurs. The results of quantum mechanics cannot be expressed in terms of classical mechanics but in abstract mathematical formulations based on complex numbers, divided into real and imaginary parts

According to quantum theory, the space of our universe is described by the real part of complex numbers, while elementary particles exist in quantum spaces explained by the imaginary part. Quantum spaces lack direct correspondence with the universe's physical space, making them inaccessible as they are not *real*. In our universe, particles appear as mathematical points—dimensionless singularities—surrounded by a field. Particles are believed to have a structure unlike our familiar physical space, filled only with energy, momentum, angular momentum, and discrete quantum numbers.

The wave function, a mathematical equation, provides information about a particle's probability amplitude, position, momentum, and other properties when measured on our side. However, the existence of these particles on their own side is problematic.

This disconnect from our classical view has troubled quantum physicists for decades. Quantum theory cannot answer where these spaces are located *physically* outside their abstract mathematical realm. Do elementary particles have a real and independent existence in these imaginary quantum spaces, beyond their probability fields?

As we understand quantum spaces, based on complex numbers, it seems that matter in our physical world is not truly physical. *It is all imaginary.*

The challenge in accepting this is that we are also composed of these *imaginary* particles.

Determining whether our bodies and souls are real, rather than mere conglomerates of mathematical points, is intriguing.

How is this possible?

Imaginary and complex numbers are widely used in scientific fields, including defining quantum spaces algebraically. Despite their popularity, scientists have noted issues with these numbers. In quantum mechanics, we struggle to relate quantum spaces to the universe's physical space.

Logic dictates that physical reality must be coherent and entirely real. Discussing holomorphicity in classical spacetime, Sir Roger Penrose highlights:[85]

> *""... the role of complex numbers in quantum theory had long struck me as a quite crucial one. If the correct geometry for the world is to be a closely quantum one, then these same complex numbers must be an essential part of this geometry. My training as a (largely pure) mathematician had taught me something of the power, subtlety and elegance of complex geometry. It had seemed fitting that this might be the geometry most basic to the structure of the physical world. Yet in its most obvious manifestations, physical geometry seems to be geometry over* $\mathbb{R}$ *(real numbers), not* $\mathbb{C}$ *(complex numbers)."*

Sir Roger Penrose's remark reveals a fundamental inconsistency. Complex numbers are efficient tools for accessing reality's properties, suggesting their algebra is suitable for interpreting reality.

However, reality should be entirely real, regardless of mathematical theories. We cannot envision even a partially imaginary reality.

This suggests our interpretation of complex numbers as having real and imaginary parts may be incorrect when applied to physical reality.

String theory seeks to provide an explanation. It suggests that these matter spaces are not hidden but have dimensions curled into small loops—compactified in mathematical terms. However, many physicists argue that string theory is ill-defined and based on crude assumptions. It is challenging to imagine that everything in the universe is composed of compactified, and thus invisible, matter.

We must consider the possibility that our understanding of the imaginary part $\mathbf{i}$ of complex numbers may be incorrect, as its interpretation varies with application. This mixed situation arises not only in quantum mechanics but also in other physics fields where complex numbers are applied, with interpretations differing in each context.

For instance, complex numbers of unit modulus can represent the group of rotations in two dimensions. Here, there is no ambiguity about what $\mathbf{i}$ represents; it is simply a rotation by 90° anticlockwise. The mathematical property $\mathbf{i}^2 = -1$ follows, and mathematicians assert there is no mystery in the interpretation.

However, the Argand diagram, where the rotation occurs, holds more than meets the eye. In other applications, such as electromagnetism, the effectiveness of complex numbers in describing electromagnetic fields remains somewhat mysterious. Here, complex numbers represent phase, but the origin of electromagnetic phase is still unclear.

The *American Heritage Dictionary of the English Language*© provides these definitions for the word imaginary [112]:

1. Existing only in the imagination; unreal.

2. Mathematics:

    a) Of, or being the coefficient of the imaginary unit in a complex number;

    b) Of, involving, or being an imaginary number;

c) Involving only a complex number of which the real part is zero.

This explanation is clear. The word "imaginary" has two primary meanings in English: it either refers to something that exists only in our minds—something unreal—or is used in mathematics to denote the imaginary part of a complex number. These definitions theoretically leave no room for misunderstanding. There should be no confusion between the mathematical meaning and its ordinary language homonym. Mathematicians assert that, from a mathematical perspective, the connotations of the word "imaginary" are irrelevant, as *it is the mathematical definition that counts.*

They distinguish between the facts—the mathematical rules defining the algebra of complex numbers—and the story of how the square root of -1 came to be called the *imaginary* part. In fact, instead of calling these numbers *imaginary*, mathematicians could have chosen any other name, and this choice would not affect the algebraic rules or their interpretation.

But they did not!

They selected the only word with such an unreal connotation as its common meaning. Initially, Descartes used this word with the derogatory intent to discourage the use of the square root of -1, considering it a *fake artifact of sloppy algebra.* Despite its oddness, the term *imaginary number* eventually gained acceptance among mathematicians, without causing confusion between its meanings.

However, when physicists sought to explain the physical nature of these imaginary spaces, they encountered a significant issue. In reality, they could not rely solely on the mathematical sense of the word imaginary, as mathematicians do. The ambiguity from the dual meaning of *imaginary* led to the common meaning overshadowing its mathematical interpretation.

Consequently, we now perceive *mathematically imaginary* numbers and spaces as *physically imaginary.* This leads us to question even the reality of our own bodies and minds. Yet, we must remember that this *extension of imaginary to the real world* has a solid foundation:

Regardless of how we label or envision quantum spaces, elementary particles appear in our universe as dimensionless singularities—or mathematical points—for which we cannot simultaneously determine both position and momentum.

One of the first scientists to address the physical interpretation of the two parts of complex numbers and resolve the ambiguity of the word imaginary was *Sir William Rowan Hamilton* (1805-1865), an Irish mathematician, Andrews Professor of Astronomy at Trinity College Dublin, and Royal Astronomer of Ireland at Dunsink Observatory [55]. He made significant contributions to optics, classical mechanics, and abstract algebra. Hamilton discovered algebraic couples, used as ordered real pairs for each real and imaginary part.

The "ordered pairs" or "algebraic couples" indicate that both parts of complex numbers can be real, existing in two different realities—or spaces in multispace [56]. The imaginary unit is no longer necessary. However, the algebra of complex numbers remains unchanged, serving as a valuable tool to represent a multispace reality. Adapting complex numbers to meet these requirements is the initial challenge. A greater challenge lies in explaining the location of these quantum spaces in our physical world and why they remain unseen.

We don't need to reinvent the wheel to address these challenges. We should follow the insights of brilliant mathematicians and physicists who have already tackled these issues. These scientists were ahead of their time, and their discoveries were not further developed or utilized by their peers.

Our first step is to revisit the history of complex numbers. The gap between quantum and classical mechanics suggests that our misunderstanding of reality might be explained not by a complex physics theory, but by a simple, fundamental concept that has been overlooked by scientists.

Let us revisit the history of complex numbers.

* * *

8 HISTORY OF IMAGINARY NUMBERS

All such expressions as $\sqrt{-1}$, $\sqrt{-2}$, $\sqrt{-3}$, $\sqrt{-4}$, etc., are impossible or imaginary numbers, since they represent roots of negative quantities; and of such numbers we may truly assert that they are neither nothing, nor greater than nothing, nor less than nothing, which necessarily constitutes them imaginary or impossible.
But notwithstanding all this, these numbers present themselves to the mind; they exist in our imagination, and we still have a sufficient idea of them.
—Leonhard Euler[42]

The earliest reference to a problem leading to imaginary numbers is found in *Stereometrica*, a treatise written by the Greek mathematician and inventor Heron of Alexandria (c. 50 AD). Heron was attempting to find the volume of a frustum (a pyramidal cross-section).

While he did not solve his problem, Heron was not the only geometer of his era to run across the square roots of negative numbers.

8.1 GIROLAMO CARDANO

Investigation into what later turned out to be complex numbers became more common in the 16th century, when the Italian mathematicians Niccolo Fontana (Tartaglia) and Girolamo Cardano began considering the roots of third- and fourth-order polynomials. They realized that even if one was only interested in real solutions to these formulas, finding them sometimes required the manipulation of square roots of negative numbers.

In 1545 Cardano published *Ars Magna* ("The Great Art"), a 40-chapter masterpiece in which he describes algebraic procedures for solving any cubic or quartic equation. He also proposed a new class of problem that deals with quadratic equations:

> *"If someone says to you, divide 10 into two parts, one of which multiplied into the other shall produce ...40, it is evident that this case or question is impossible. Nevertheless, we shall solve it in this fashion."*

To solve the problem, Cardano applied the high-school method of completing the square. Combining $x + y = 10$ with $xy = 40$ yields the quadratic equation $x^2 - 10x + 40 = 0$, which has the solutions $5 \pm \sqrt{-15}$. Cardano obtained this result, but dismissed it as an invalid solution.

In connection with cubic equations, Cardano's formula gives solutions to equations of the form $x^3 = ax + b$ as follows:

$$x = \sqrt[3]{\frac{b}{2} + \sqrt{\left(\frac{b}{2}\right)^2 - \left(\frac{a}{3}\right)^3}} + \sqrt[3]{\frac{b}{2} - \sqrt{\left(\frac{b}{2}\right)^2 - \left(\frac{a}{3}\right)^3}}.$$

(8.1)

When used to solve the classic example

$$x^3 - 15x - 4 = 0, \tag{8.2}$$

the formula yields

$$x = \sqrt[3]{2 + \sqrt{-121}} + \sqrt[3]{2 - \sqrt{-121}}. \tag{8.3}$$

Cardano claimed that the general formula was not applicable in this case, because the square root of -121 could not be calculated. Unfortunately, his geometric mindset made it hard for him to make further headway on complex numbers. At one point in *Ars Magna,* he comments that arithmetic dealing with quantities such as $\sqrt{-1}$ *involves mental tortures and is truly sophisticated.* He held a similar view of his solution to the unrealistic quadratic equation, which he dubbed *as refined as it is useless.*

Many mathematicians shared his feelings, but eventually there was a breakthrough.

8.2 RAFAEL BOMBELLI

In 1572, the hydraulic engineer Rafael Bombelli (1526-1572) took up the challenge. He published *L'Algebra,* a set of three books on the theory of polynomials that (he claimed) *would enable a novice to master the subject.*

The opus became a classic of mathematical literature. Leibnitz cited Bombelli as an outstanding master of the analytical arts.

Bombelli justified Cardano's cubic equation $x^3 - 15x - 4 = 0$ (8.2) via a detour by using the terms *più di meno*

L'ALGEBRA
OPERA
Di Rafael Bombelli da Bologna
Diuisa in tre Libri.
Con la quale ciascuno da se potrà venire in perfetta cognitione della teorica dell' Arimetica.
Con vna Tauola copiosa delle materie, che in essa si contengono.
Posta hora in luce à beneficio delli studiosi di detta professione.
IN BOLOGNA,
Per Giouanni Rossi. MDLXXIX.
Con licenza de' Superiori.

Figure 8.1: The Bombelli's Algebra book

(*pdm*) and *meno di meno* (*mdm*) to represent what later became $+\mathbf{i} = +\sqrt{-1}$ and $-\mathbf{i} = -\sqrt{-1}$ respectively. He did that by introducing a solution of the form $x = (2 + \mathbf{i}) + (2 - \mathbf{i})$.

Simple arithmetic then gives $x = (2+\mathbf{i}) + (2-\mathbf{i}) = 4$, and it could easily be checked that $x = 4$ is a correct solution to the original equation. However, Bombelli found it only by taking a very clever detour through the uncharted territory of complex numbers. He basically assumed that numbers of the form

$$a + \sqrt{-b} \tag{8.4}$$

existed, and applied the normal operational rules of algebra. In this particular case, Bombelli's insight was that since the radicands

$$2+\sqrt{-121} \quad \text{and} \quad 2-\sqrt{-121} \tag{8.5}$$

differed only in their sign, the similar rule might apply to their cube roots. He thus wrote

$$2+\sqrt{-121}=a+\sqrt{-b}, \tag{8.6}$$

and

$$2-\sqrt{-121}=a-\sqrt{-b}, \tag{8.7}$$

and went about solving for a and b. The result was surprisingly simple: $a=2$ and $b=1$. Thus,

$$\begin{aligned} x &= \sqrt[3]{2+\sqrt{-121}}+\sqrt[3]{2-\sqrt{-121}} \\ &= \left(2+\sqrt{-1}\right)\left(2-\sqrt{-1}\right)=4. \end{aligned} \tag{8.8}$$

This method gave the correct solution to the equation, and convinced Bombelli that his ideas had some validity.

Bombelli also worked out a formal algebra of complex numbers. In the first book of *L'Algebra,* he uses the terms *più di meno (pdm)* and *meno di meno (mdm)* to represent $+\mathbf{i}$ and $-\mathbf{i}$ respectively, then gives eight *basic* rules for calculating with these entities, establishing the fundaments of the complex number algebra[9]. We added their modern notation:

Più via più di meno, fa più di meno: $+1\cdot+\mathbf{i}=+\mathbf{i}$

Più via meno di meno, fa meno di meno: $+1\cdot-\mathbf{i}=-\mathbf{i}$

Più di meno via più di meno, fa meno: $+\mathbf{i}\cdot+\mathbf{i}=-1$

Meno di meno via più di meno, fa più: $-\mathbf{i} \cdot +\mathbf{i} = +1$

Meno via più di meno, fa meno di meno: $-1 \cdot +\mathbf{i} = -\mathbf{i}$

Meno via meno di meno, fa più di meno: $-1 \cdot -\mathbf{i} = +\mathbf{i}$

Più di meno via meno di meno, fa più: $+\mathbf{i} \cdot -\mathbf{i} = +1$

Meno di meno via meno di meno, fa meno: $-\mathbf{i} \cdot -\mathbf{i} = -1$

In his book Algebra (1572), Rafael Bombelli defined numbers like $\sqrt{-1}$ by giving them special names and specifying arithmetic rules distinct from those for real numbers. He called $\sqrt{-1}$ "plus of minus" and $-\sqrt{-1}$ "minus of minus," making it clear these were neither positive nor negative and should not be treated as ordinary radicals. However, nobody knows what he privately understood about their signification. The fact is that his algebra is too precise and complex to be merely a product of random thought. It is likely that he imagined a second, orthogonal space, but he explained $\sqrt{-1}$ as a *wild thought* to avoid appearing foolish.

According to A. Sfard, this artifact was introduced solely as an operational concept, but this was his supposition about Bombelli's thinking:

> *"The symbol $\sqrt{-1}$ was initially considered nothing more than an abbreviation for certain "meaningless" numerical operations. It came to designate a fully fledged mathematical object only after mathematicians got accustomed to these strange but useful kinds of computation [102]."*

8.3 RENÉ DESCARTES

Many mathematicians did not make this distinction and considered complex numbers a waste of time.

Descartes, a French philosopher, mathematician, and writer, wrote *La Géométrie* in 1637[34], where he viewed square roots of negative numbers as geometric impossibilities.

René Descartes (1596-1650) coined the term *imaginary numbers* with a derogatory intent. He noted that while one might imagine an equation having as many roots as its degree, not all imagined roots correspond to real numbers.

Descartes saw these numbers as *fake artifacts of sloppy algebra*, commenting that *even though one can imagine that an equation has as many roots as its degree, there might not exist real numbers corresponding to all of the imagined roots.*

A source of confusion was the equation

$$\sqrt{-1}^2 = \sqrt{-1}\sqrt{-1} = -1 \tag{8.9}$$

which seemed inconsistent with the algebraic identity

$$\sqrt{a}\sqrt{b} = \sqrt{ab}. \tag{8.10}$$

This relation was valid for all positive real numbers a and b, and could be used when one number was positive and the other negative. The incorrect application of this identity (and its related identity $\frac{1}{\sqrt{a}} = \sqrt{\frac{1}{a}}$) to cases where *both* a and b are negative troubled even Euler.

Despite Descartes using the term *imaginary* derogatorily to dismiss square roots of -1, it persisted over centuries, subtly hindering scientific progress.

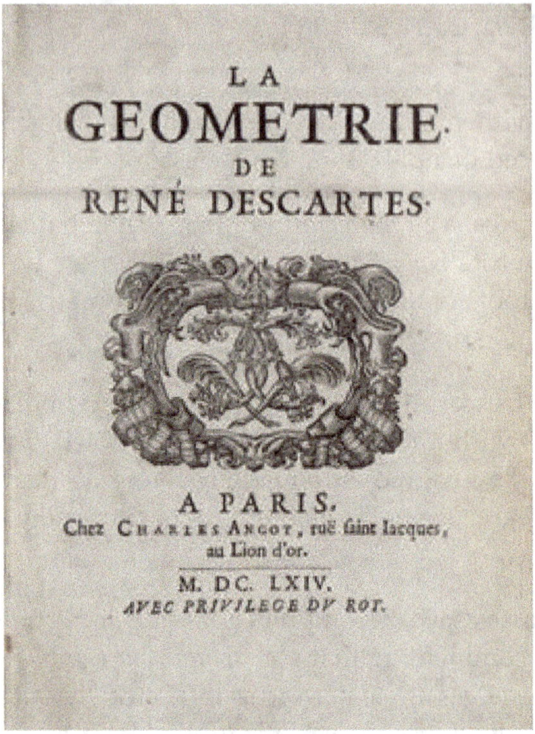

LA
GEOMETRIE.
DE
RENÉ DESCARTES.

A PARIS,
Chez CHARLES ANGOT, ruë saint Iacques,
au Lion d'or.
M. DC. LXIV.
AVEC PRIVILEGE DV ROY.

Figure 8.2: La Géométrie book

The difficulty of such calculations led to the creation of the symbol **i** for $\sqrt{-1}$; this convention helped prevent mistakes.

Gottfried Wilhelm von Leibniz (1646-1716) spent considerable time applying algebraic laws to complex numbers. Both he and Johann Bernoulli used imaginary numbers as integration aids. Johann Lambert used complex numbers for map projection, and Jean D'Alembert applied them in hydrodynamics theory.

Two centuries later, Carl Friedrich Gauss (1777-1855), a German mathematician, made extensive contributions to several scientific fields. Considered one of the *greatest mathematicians since antiquity,* Gauss significantly influenced mathematics, physics, and astronomy, ranking as one of history's most influential mathematicians.

Gauss clarified the concept of *complex numbers* but criticized the term [46]:

> *"That this subject [of imaginary magnitudes] has hitherto been considered from the wrong point of view and surrounded by a mysterious obscurity, [it] is to be attributed largely to an ill-adapted notation. If, for example,* $+1, -1, \sqrt{-1}$ *had been called direct, inverse and lateral units, instead of positive, negative and imaginary (or even impossible), such an obscurity would have been out of the question."*

However, he could not persuade his peers to abandon the troublesome *imaginary* term, and he was not alone in attempting this change.

8.4 LEONHARD EULER

In the 18th century, Swiss mathematician and physicist Leonhard Euler (1707-1783) contributed significantly to mathematics and physics, introducing several modern concepts and notations, including the letter i for the square root of -1.

He proposed

$$a + b\mathrm{i}$$

as the general form of a complex number.

In 1748, Euler published the renowned *Euler's formula*, widely used in modern science to model waves and oscillations across various media [73]. This formula helped complex numbers gain acceptance among mathematicians.

In 1765, Euler published *Elements of Algebra*, providing a comprehensive introduction to elementary algebra, including a discussion on the nature of numbers [42].

Two chapters, *12. Of Square Roots, and of Irrational Numbers resulting from them*, and *13. Of Impossible, or Imaginary Quantities, which arise from the same source*, elucidate the distinct nature of these numbers. Euler emphasizes the unique role of the imaginary unit:

> *"Now, as $-a$ is equal to $+a$ multiplied by -1, and as the square root of a product is found by multiplying together the roots of its factors, it follows that the root of a multiplied by -1, or $\sqrt{a}$, is equal to $\sqrt{a}$ multiplied by $\sqrt{-1}$. Now $\sqrt{a}$ is a possible or real number, consequently* the whole impossibility of an imaginary quantity may be always reduced to $\sqrt{-1}$. *For this reason, $\sqrt{-4}$ is equal to $\sqrt{4}$ multiplied by $\sqrt{-1}$ and equal to $2\sqrt{-1}$, on account of $\sqrt{4}$ being equal to 2.*
> *For the same reason $\sqrt{-9}$ is reduced to $\sqrt{9} \cdot \sqrt{-1}$, or $3\sqrt{-1}$, and $\sqrt{-16}$ is equal to $4\sqrt{-1}$."*

It is crucial to note Euler's insightful observation:

> *"The whole impossibility of an imaginary quantity may be always reduced to $\sqrt{-1}$."*

In 1777, Euler introduced the notation $\mathbf{i}$ and $-\mathbf{i}$ for the two distinct square roots of -1. He also developed the notation $a + b\mathbf{i}$ for a general complex number. Euler then began to investigate the behavior of functions beyond polynomials with complex-valued arguments, such as the exponential function. His work led to one of the most renowned equations in mathematics, as revolutionary and profound as Einstein's $E = mc^2$:

$$\cos\theta + \mathbf{i}\sin\theta = e^{\mathbf{i}\theta}, \tag{8.11}$$

which for the special case $\theta = \pi$ results in

$$e^{\mathbf{i}\pi} = -1. \tag{8.12}$$

The Euler formula (1748) is predominantly used in modern science to model waves and oscillations, regardless of their nature or medium. For mathematicians, the identity obtained by choosing $\theta = \pi$ is remarkable as it connects three fundamental real numbers: Euler's constant e, π, and the number 1. No prior relationship between these had been discovered; it is akin to physicists finding a new law linking the gravitational constant to the electron's charge [42].

The 18th century also saw contributions from Abraham de Moivre, whose work inspired Euler. This is evident in the well-known formula named after him (1730):

$$(\cos\theta + \mathbf{i}\sin\theta)^n = \cos n\theta + \mathbf{i}\sin n\theta. \tag{8.13}$$

De Moivre's formula enabled mathematicians to raise complex numbers to any power by describing the behavior of a unit vector on the complex plane. This result completed the extension of standard algebraic operations to complex numbers, establishing complex analysis as a unique and consistent mathematical theory.

8.5 CASPAR WESSEL

Despite Euler's work, complex numbers were not fully accepted until Caspar Wessel (1745-1818), a Norwegian-Danish mathematician, published a geometrical interpretation in 1799.

Wessel proposed representing complex numbers on a modified Cartesian plane, with the horizontal line as the real number axis (unit 1) and the vertical line as the imaginary number axis (unit **i**).

A complex number $z = a + b\mathbf{i}$ can be depicted as a point or vector in the complex plane:

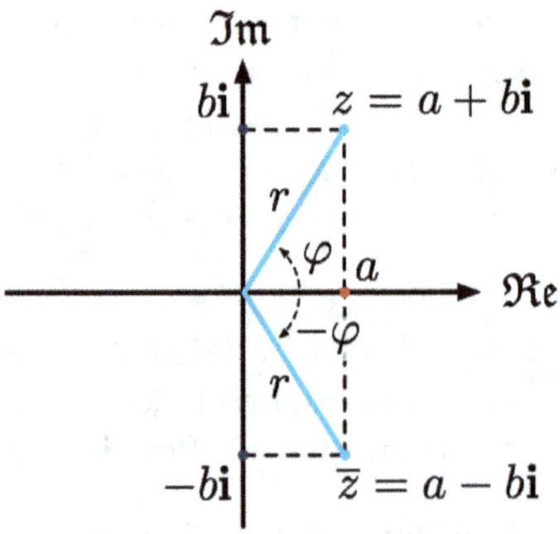

Figure 8.3: The Argand diagram

The numbers a and b are real, as in the Cartesian plane, but b, on the imaginary axis, is multiplied by the imaginary unit **i**. Geometric representations of complex numbers are known as *Argand diagrams* or *Argand planes*.

Complex analysis results became more intuitive when viewed through this lens. Wessel's work gained popularity through Johann Carl Friedrich Gauss in the nineteenth century, significantly expanding interest in complex number theory.

It's noteworthy that Wessel wasn't the only one to propose the complex plane; similar concepts emerged in 1685 (John Wallis), 1804 (Abbé Buée), 1806 (Jean-Robert Argand), and 1828 (Warren and Mourey). Although Wessel's paper was published first, historical analysis credits Argand with an independent discovery.

Geometric representations of complex numbers are still referred to as Argand diagrams. Wessel's description of the complex plane appeared in *Proceedings of the Copenhagen Academy* (1799). In the same paper, he used William Hamilton's quaternion theory (1843) to develop spherical surface trigonometry.

The complex plane visualization helped *imaginary* and *complex* numbers gain acceptance in mainstream mathematics as a natural extension of negative numbers[123].

Augustin Louis Cauchy and Niels Henrik Abel also achieved great success with complex analysis, encouraging others to enter the field. In analogy to vectors, Argand called $\cos\phi + \mathbf{i}\sin\phi$ the *direction factor*, and $r = \sqrt{a^2 + b^2}$ the *modulus* (also known as the amplitude or norm in modern terminology).

Augustin-Louis Cauchy[17] (1789-1857) was a French mathematician who founded modern complex function theory and made significant contributions to abstract algebra. In his book *Cours d'Analyse*, Cauchy emphasized rigorous analysis, and in 1847, he dismissed the use of the imaginary unit.

> *"We can completely repudiate the symbol $\sqrt{-1}$, abandoning it without regret because one does not know what this pretended sign signifies nor what sense one ought to attribute to it."*

In an article published in 1874 in the *Proceedings of the American Academy of Arts and Sciences,* Benjamin Peirce presents an intriguing perspective:

> *"When the formulas admit of intelligible interpretation, they are accessions to knowledge... But the most noted instance is the symbol $\sqrt{-1}$, called the* impossible *or* imaginary, *and which... may be more definitely distinguished as the symbol of* semi-inversion... *The strongest use of the symbol is to be found in its* magical power of doubling the actual universe, and placing by its side an ideal universe, its exact counterpart, with which it can be compared and contrasted, *and, by means of curiously connecting fibres, form with it an organic whole, from which modern analysis has developed her surpassing geometry [50]."*

Like other eminent mathematicians who believed the *imaginary* concept should be replaced with a more *real* one, Peirce suggests an *ideal universe,* positioned alongside the real universe, as *its exact counterpart.*

Gauss published the first *correct* proof of the fundamental theorem of algebra in his 1797 doctoral thesis, yet by 1825, he still claimed that *the true metaphysics of the square root of -1 is elusive.* By 1831, however, Gauss had resolved some of his reservations about complex numbers and published works on their geometric representation.

Gauss's student Ferdinand Eisenstein studied numbers of the form $a + b\omega$, where ω is any complex root of the polynomial $x^3 - 1$. The theory of complex numbers and their variants has continued to evolve. Besides their crucial role in proving the fundamental theorem of algebra, they have been invaluable in solving numerous problems in geometry and number theory [24].

8.6 WILLIAM ROWAN HAMILTON

The scientist who made the most significant contribution to understanding *imaginary numbers* was William Rowan Hamilton (1805-1865), an Irish mathematician, physicist, and astronomer. He introduced new mathematical concepts and techniques in classical mechanics, electromagnetism, and optics. In pure mathematics, he is renowned for discovering quaternions, an algebraic tool used in quantum mechanics.

Hamilton aimed to simplify working with imaginary quantities by expanding *real* numbers to a larger set he called *algebraic couples*, denoted as (a, b), where both a and b are real numbers. (Today, *algebraic couples* are known as *ordered pairs*.)

In his book *Theory of Conjugate Functions, or Algebraic Couples; with a Preliminary and Elementary Essay on Algebra as the Science of Pure Time*,[55] published in 1837, Hamilton wrote:

> *"The* Theory of Conjugate Functions *gives reality and meaning to conceptions that were before* Imaginary, Impossible, or Contradictory... *The*

> *author acknowledges with pleasure that he agrees with M. Cauchy, in considering every (so-called)* Imaginary Equation *as a symbolic representation of two separate Real Equations: but he differs from that excellent mathematician... in not introducing the sign* $\sqrt{-1}$ *until he has provided for it, by his Theory of Couples, a possible and real meaning, as a symbol of the couple* $(0,1)$.*"*

Mathematically, Hamilton achieved his goal. He explains:

> *"In the theory of single numbers, the symbol* $\sqrt{-1}$ *is* absurd, *and denotes an* impossible extraction, *or a merely* imaginary number; *but in the* theory of couples, *the same symbol* $\sqrt{-1}$ *is significant, and denotes a* possible extraction, *or a* real couple..."

Hamilton's *Theory of Couples* posits that the imaginary unit $\sqrt{-1}$ serves two purposes. Firstly, it identifies the secondary unit $(0,1) = \sqrt{-1}$ in the ordered pair $(a,b) = a + b\sqrt{-1}$, where a and b are real numbers. Secondly, while the powers of the primary unit $(1,0)$ remain $(1,0)$, the powers of the secondary unit $(0,1)$ cycle through $(0,1), (-1,0)$,
$(0,-1), (1,0)$, illustrating its dynamic nature.

This property of the secondary unit, the imaginary unit **i** in complex numbers, underscores the importance of examining the history of complex numbers.

We note that three centuries later, Hamilton rediscovered Bombelli's concept of a secondary space alongside our universe.

8.7 ARE IMAGINARIES REAL?

Imaginary numbers, defined by mathematical rules, should theoretically be interpreted without bias. However, debates about their nature have persisted since their inception.

Some critics argue that ... *mathematicians belong with poets and theologians as mystics,* because they believe in *numbers that have no reality* [59].

Many physicists view mathematical models as tools without deeper meaning, not indicative of reality. Conversely, the realist perspective suggests that mathematical elements of successful theories hint at underlying reality. If they don't reflect reality, the theory hasn't truly expanded our knowledge.

Thus, a meaningful representation of imaginary and complex numbers should eventually align with the physical world. In current science, they effectively describe electromagnetic and quantum fields.

As Penrose noted, *physical geometry, in its most apparent forms, appears to be geometry over* $\mathbb{R}$ *(real numbers), not* $\mathbb{C}$ *(complex numbers)*[86]. Thus, the physics of imaginary spaces seems illogical.

However, Penrose's paradox suggests a plausible explanation:

> *"Imaginary numbers and spaces are in fact real,*
> *but we just do not understand them correctly yet."*

However, this change in multi-space reality.

It is time to reveal the true nature of *imaginary numbers and spaces*.

* * *

9 PHYSICAL NATURE OF IMAGINARIES

> *Einstein said that behind every great theory there is a simple physical picture that even a child could understand. In fact, he said, if a theory does not have a simple underlying picture, then the theory is probably worthless.The important thing is the physical picture; math is nothing but bookkeeping.*
> —Michio Kaku

Considering the above citation, this chapter aims to present a *simple physical picture* of complex numbers theory, eliminating the enigmatic imaginary world at reality's core. We assume readers are familiar with the basic mathematical rules of complex numbers and can refresh their knowledge via Wikipedia or other sources. For simplicity, some assumptions or explanations may not withstand rigorous scrutiny by purists, as we are exploring a superficially chartered territory.

As Einstein stated: *The important thing is the physical picture.*

9.1 COMPLEX NUMBER DEFINITION

A complex number $z = a + b\mathrm{i}$ consists of three components: two real numbers a and b, and the imaginary unit

$\mathbf{i}$. In z, a is the real part, and b in $b\mathbf{i}$ is the imaginary part. Conventionally, the imaginary part excludes the imaginary unit $\mathbf{i}$.

The complex plane, introduced in a previous chapter in Figure 8.3 on page 78, is another convention defined by mathematical rules. Einstein remarked that *math is nothing but bookkeeping.* He is not alone in this view. Many physicists believe mathematical models of observations lack deeper meaning and their elements should not be seen as indicative of reality.

While respecting their opinion, I support the realist viewpoint that the mathematical elements of successful theories hint at the nature of underlying reality. A correct representation of complex numbers should ultimately make sense in the physical world. If the central mathematical elements of a theory do not reflect reality, then the theory has not expanded our knowledge; it has only described what we already know or can observe.

9.2 REAL NUMBERS ON A COMPLEX PLANE

In complex numbers, the only difference between a and b is their numerical values. In Figure 9.1 below, we can represent them on axis $\mathfrak{Re}$ of the complex plane.

Multiplying b by $\mathbf{i}$ (in fact by $+\mathbf{i}$) labels it as the *imaginary part* and moves it on axis $\mathfrak{Im}$, as shown in Figure 9.2. Interestingly, reversing the operation by multiplying $b\mathbf{i}$ by $-\mathbf{i}$ brings it back to the real axis: $b\mathbf{i} \cdot -\mathbf{i} = b$

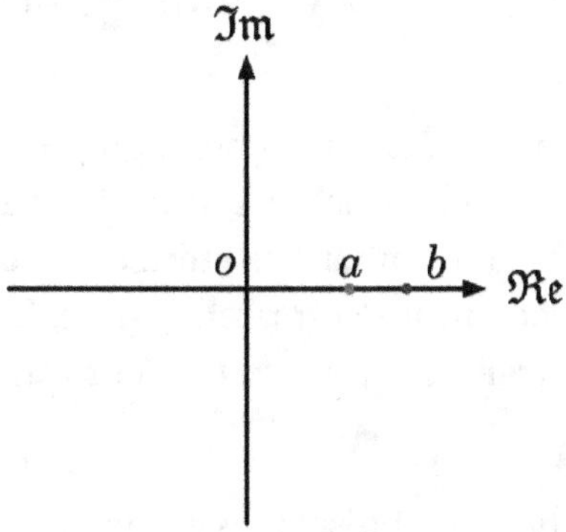

Figure 9.1: The real numbers a and b on a complex plane

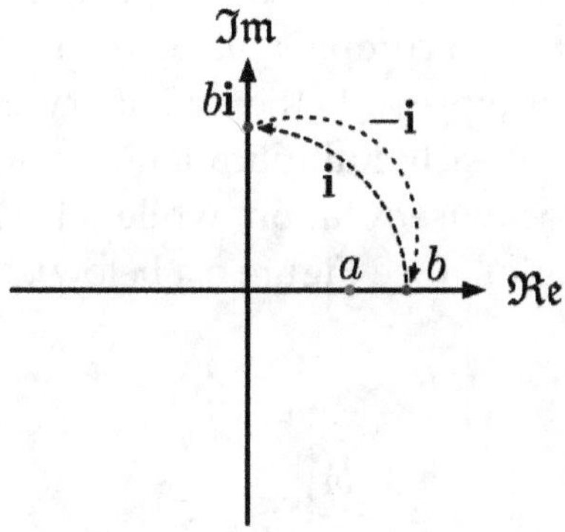

Figure 9.2: Numbers a and b are real, but on different axes

The debate over the nature of *imaginary* numbers emerged alongside the numbers themselves. Multiplying a number from the *real* space $\mathfrak{Re}$ by $\mathbf{i}$ to move it to the *imaginary* space $\mathfrak{Im}$ initially seems odd.

9.3 NATURE OF IMAGINARY UNIT

However, b's intrinsic nature—belonging to the set of real numbers $\mathbb{R}$—remains unchanged on the imaginary axis. By stating that the number has shifted its *community*, this concept becomes more comprehensible: The imaginary unit $\mathbf{i}$ simply transfers b from one community (axis) to the other.

Alternatively, $\mathbf{i}$ can be seen as indicating that b is now on axis $\mathfrak{Im}$ instead of $\mathfrak{Re}$. Viewing each axis as a unidimensional space, $+\mathbf{i}$ shifts b to an imaginary space, while $-\mathbf{i}$ returns it to the real space.

In the complex plane, $\mathbf{i}$ does more than move a number from $\mathfrak{Re}$ to $\mathfrak{Im}$. In a complex plane, it has two geometric functions: (1) it represents the imaginary unit of the $\mathfrak{Im}$-axis [66], and (2) each multiplication of b by $\mathbf{i}$ results in a 90° counterclockwise rotation, while $-\mathbf{i}$ results in a 90° clockwise rotation. (See Figure 9.3 below).

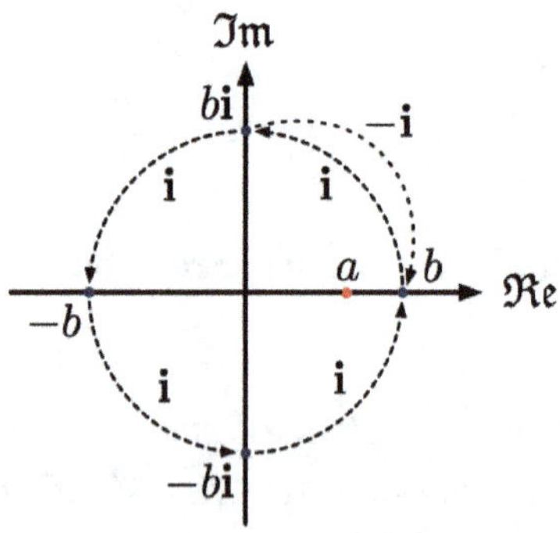

Figure 9.3: Multiplying a number by $\mathbf{i}$ or -$\mathbf{i}$ rotates it by 90°

The effect of two consecutive left or right turns is to reverse direction, creating a *U*-turn. The statement $\mathbf{i}^2 = -1$ algebraically represents the geometric fact that two quarter turns, $90° + 90°$, equal a straight angle of $180°$.

The integer powers of **i** follow a modulo-four cycle:

$$\mathbf{i}^0 = 1,\ \mathbf{i}^1 = \mathbf{i},\ \mathbf{i}^2 = -1,\ \mathbf{i}^3 = -i,\ \mathbf{i}^4 = 1.$$

This function of the imaginary unit **i** leads to its common designation as the **i**-operator (or **j**-operator in electrical engineering).

With each quarter turn, the number b shifts from one axis ($\Re$ or $\Im$) to the other, yet b remains a real number.

9.4 TWIN COMMUNITIES

This discussion has focused on the integer powers of **i** to better understand the *twin communities* within complex numbers. This focus does not imply any disregard for Euler's formula introduced earlier.

Recall that A. Sfard noted Bombelli's introduction of $\sqrt{-1}$ as a *meaningless numerical operation.*

Three centuries later (1874 vs 1572), Hamilton recognized the significance of these *twin communities*, stating that *while in the theory of single numbers, the symbol* $\sqrt{-1}$ *is absurd and denotes... a merely imaginary number... in the theory of couples, the same symbol* $\sqrt{-1}$ *is significant and denotes... a real couple...*

In Hamilton's algebraic couples, the powers of the secondary unit $(0, 1)$ correspond to the integer powers of i:

$$
\begin{aligned}
\mathrm{i}^0 &= 1 = (1,0), \\
\mathrm{i}^1 &= \mathrm{i} = (0,1), \\
\mathrm{i}^2 &= -1 = (-1,0), \\
\mathrm{i}^3 &= -\mathrm{i} = (0,-1), \\
\mathrm{i}^4 &= 1 = (1,0).
\end{aligned}
$$

However, constructing complex numbers as algebraic couples replaces the imaginary concept with the notion of ordered pairs of real numbers, which are already meaningful in physical reality.

Let us trace the evolution of this concept.

9.5 CARTESIAN GEOMETRY

In 1637, Descartes developed Cartesian, or analytic geometry, using algebra to describe geometry.

In 1799, Wessel reversed this approach: he used geometry to represent the algebra of complex numbers.

He constructed a modified Cartesian plane, where the horizontal line is the axis of real numbers, and the vertical line is the axis of imaginary numbers. Thus, a complex number $z = a + b\mathrm{i}$ is depicted as a point (a, b) or vector in the complex plane.

In 1833, Hamilton introduced algebraic couples (a, b), revisiting the Cartesian representation of ordered pairs (a, b).

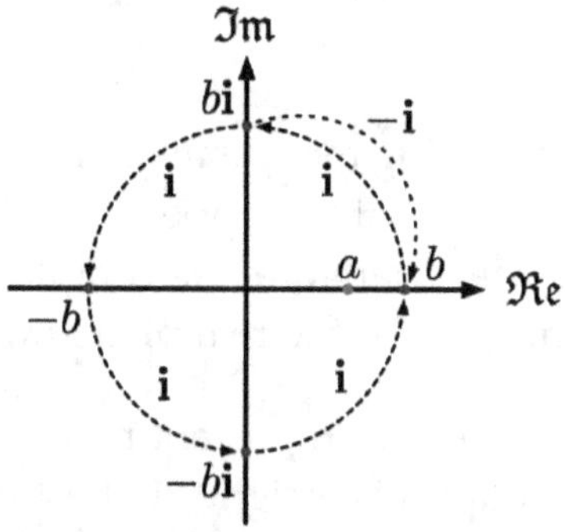

Figure 9.4: Cartesian coordinate system

9.6 OPERATOR FUNCTION OF I

Both coordinate systems use ordered pairs of real numbers, yet they differ from Descartes' original concept. In the real plane, the powers of the primary and secondary units $(1,0)$ and $(0,1)$ remain equal to the units themselves. However, in the complex plane, the y-axis is deemed *imaginary* because the successive powers of the secondary unit (0, 1) cause it to *oscillate between the two sides of the couple:* $(0,1),(-1,0),(0,-1),(1,0)$.

If we temporarily disregard the *operator function of* **i**, both axes of the plane for algebraic couples are as real as those for ordered pairs. Thus, the initial function of **i** to represent the imaginary unit of the ℑm-axis vanishes: there is no imaginary axis, as both axes are real. This is well recognized in physics and engineering, where complex algebra is frequently used for mathematical operations within real spaces. This suggests that the **i**-*operator* is not inherently linked to the physical nature of the two spaces.

However, considering the *operator function of* **i** and viewing the plane as a whole, it appears similar yet *distinct* from real 2D space. Therefore, the complex plane (a, b) is said to be isomorphic to the real plane (a, b).

Hamilton's concept of representing complex numbers with algebraic couples is insightful: it establishes that the complex plane is a two-dimensional coordinate system, both dimensions labeled as *real*. The distinction from the Cartesian plane arises from the relationship between the two axes, where one dimension appears *imaginary* relative to the other.

In physics, this conclusion suggests we should consider quantum spaces as real entities, with the understanding that we must comprehend their *imaginary* relationship physically, not just mathematically.

To achieve this, we must identify a physical spatial configuration that maintains isomorphicity with the Cartesian plane while involving two real and independent spaces. This requires discussing two well-known physics concepts that were not previously linked to complex spaces.

9.7 FRAMES AND OBSERVERS

The first concept is *reference frames* (or *frames of reference*), which arose from the need to observe phenomena from specific viewpoints, whether by humans or measurement devices, collectively called *observers*.

For instance, in mathematics, the Cartesian plane is an *abstract 2D coordinate system.*

In physics, *coordinate systems* remain abstract but are tied to *frames of reference,* sets of physical reference points that uniquely locate and orient them. Thus, the mathematical *Cartesian plane* is part of a physical *Cartesian frame of reference.*

The second concept is the *observer* $\mathbb{O}$, crucial in relativity and quantum mechanics.

Special relativity is entirely based on *observational frames of reference,* which are in relative inertial motion [72].

Their constant state of motion also labels them as *inertial frames of reference.* Essentially, each frame contains either an observer at rest or the observed phenomena.

In quantum mechanics, the observer's role includes not just measurement but also the issue of the observed phenomena. As physicist Fritjof Capra explains [115]:

> *"The crucial feature of atomic physics is that the human observer is not only necessary to observe the properties of an object, but is necessary even to define these properties... This can be illustrated with the simple case of a subatomic particle. When observing such a particle, one may choose to measure - among other quantities - the particle's position and its momentum."*

9.8 RELATIVITY OF SIMULTANEITY

Einstein discovered the special relativity of reference frames in 1905 [39].

In his theory, Einstein shows that time is not absolute and is inseparably linked to light velocity. This led to the discovery of the *relativity of simultaneity*, meaning *inertial observers in relative motion have different times.*[78]

However, Einstein did not address space. *In fact, he did not realize that two observers in relative motion, who have different times, must also have different spaces.*

In recent decades, physicists have emphasized the need to distinguish between mathematical coordinate systems and physical frames of reference. J. D. Norton provides a succinct description of this situation [77]:

> *"In traditional developments of special and general relativity it has been customary not to distinguish between two quite distinct ideas. The first is the notion of a coordinate system, understood simply as the smooth, invertible assignment of four numbers to events in spacetime neighborhoods. The second, the frame of reference, refers to a system used to assign such numbers... More recently, to negotiate the obvious ambiguities of Einstein's treatment, the notion of frame of reference has reappeared as a structure distinct from a coordinate system."*

As Patrick Cornille explains [28]:

> *"A distinction between mathematical sets of coordinates and physical frames of reference must*

> *be made. The ignorance of such distinction is the source of much confusion... the dependent functions such as velocity for example, are measured with respect to a physical reference frame, but one is free to choose any mathematical coordinate system in which the equations are specified."*

Graham Nerlich made a similar assertion [75]:

> *"The idea of a reference frame is really quite different from that of a coordinate system.*Frames differ just when they define different spaces (sets of rest points) or times (sets of simultaneous events). *(Emphasis added) So the ideas of a space, a time, of rest and simultaneity, go inextricably together with that of frame. However, a mere shift of origin, or a purely spatial rotation of space coordinates results in a new coordinate system."*

This leads us to the hypothesis that reference frames may be more than mere mathematical coordinate systems. Perhaps, the reference frames of special relativity should be considered as independent physical spaces.

Vesselin Petkov highlights this idea when discussing Minkowski's view on what he called *many spaces* [88]:

> *"While Einstein, unlike Lorentz, insisted that the times of all inertial observers are equally good, Minkowski noticed that* neither Einstein nor Lorentz made any attack on the concept of space *and stressed that the idea of many spaces is unavoidable in special relativity [70]. "*

9.9 NO ABSOLUTE SPACE

In fact, Norton, Cornille, Nerlich, and Petkov suggested that Einstein's special theory of relativity might be flawed if reality is based on just one absolute space.

We typically think of one real space with one, two, or three dimensions. Descartes led us to believe that numbers multiplied by the square root of -1 are imaginary, but the numbers themselves are always real.

If this hypothesis is correct, we should be able to perceive these independent spaces. However, this is challenging because our senses may not distinguish these spaces from each other and the universe's space.[1]

NOTE: This interpretation of **i** is supported by the fact that, to keep tabs on operations, the **i**-*part* must always be calculated first, before numeric calculations are processed on each axis.

In mathematics, definitions are absolute and do not need an observer[60], e.g., a *real number* is *a number that is either rational or the limit of a sequence of rational numbers.*[113]

However, in several fields of physics, e.g., in special and general relativity and quantum mechanics, the observer's role is mandatory. The seemingly different natures of its

---

1 If you think it is impossible for us not to differentiate one such space from another, remember that our senses are far from capable of acquiring all the information available in the surrounding world. For example, our bodies cannot directly experience "speed." To validate this assertion, recall that the average orbital speed of the Earth around the Sun is about 30 km/s (108,000 km/h). So we need strong supporting evidence to prove the hypothesis wrong.

two coordinates, *real* $\mathfrak{Re}$ and *imaginary* $\mathfrak{Im}$, suggest that the Argand space could be understood as a reunion of two independent 1D vector spaces $\mathbb{S}_1$ and $\mathbb{S}_2$, one real and the other imaginary. These spaces are orthogonal and, in a sense, cannot perceive each other.

To avoid the entanglement between the casual meaning of the word *imaginary* and its mathematical meaning, we will use the concept of the observer from physics. Thus, we proceed by replacing the questionable names *real* and *imaginary* with two observer-related explanations.

9.10 MATHEMATICAL DUAL-SPACE

We imagine a Gedankenexperiment—as Einstein liked to call these experiments—in which there are two independent spaces $\mathbb{S}_1$ and $\mathbb{S}_2$, which cannot directly see each other—such as New York and Melbourne.

We start our inquiry by imagining that, in space $\mathbb{S}_1$, there are two observers $\mathbb{O}_1$ and $\mathbb{O}_2$, who look at two objects a and b as depicted in Figure 9.1 on page 87.

The number is:

$$z_0 = a + b \mid a, b \in \mathbb{R}$$

Seen from the perspective of the two observers, the space $\mathbb{S}_1$ is real and can be defined as the space of axis $\mathfrak{Re}$. They cannot directly experience space $\mathbb{S}_2$, which is considered the space of axis $\mathfrak{Im}$.

We designate the space S_1 of the two observers O_1 and O_2 as the *here space*; this space is real for both observers and surrounds them.[2]

With both observers in S_1, the space S_2 is remote for them and is termed the *there space*. It is also real but different and independent from the observers' space. Because the observers do not reside in this space, it is not directly accessible—it is invisible—to them, and they perceive it as *imaginary*.

By definition, observer O_1 inhabits only space S_1, but observer O_2 can travel to either of the two spaces.[3]

9.11 DUAL CONCEPT: *HERE* AND *THERE*

In general, the *here* space is by default the space of the universe, and the *there* space remains the imaginary space.

Let us redraw Figure 9.1 with the complex set of two orthogonal 1D real spaces.

We decide to send b to space S_2. To do this, we remember that **i** has the properties (1) to move a number from one axis to the other—from one space to the other—and (2) to remain attached to the number after the move, as a flag identifying its position on two of the four half-axes.

2 We normally assume that *real* is something that exists in the space of our universe, which is real by default.

3 For observers residing in space S_1, a and b are *real* because they are *here*, in the same space.

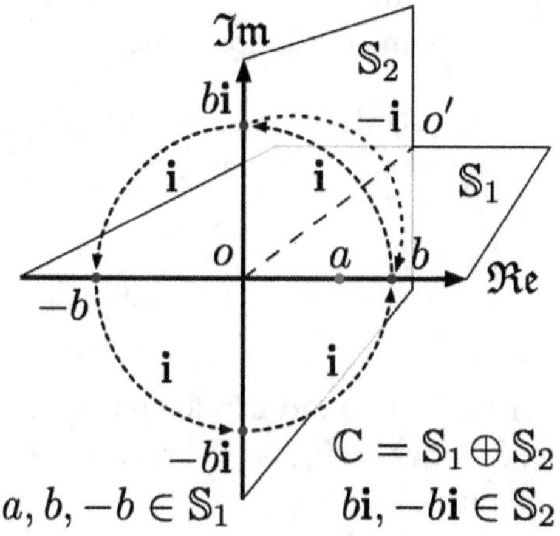

Figure 9.5: The complex set of two orthogonal 1D real spaces

We multiply b by $\mathbf{i}$, so in $\mathbb{S}_2$, b becomes $b\mathbf{i}$. Now, there is a number in each space, so instead of the number $z_0 = a + b$, the new number becomes:

$$z_1 = a + b\mathbf{i} \mid a, b \in \mathbb{R}$$

The two numbers a and b are still both real, but one (a) is *here* in $\mathbb{S}_1$ and the other (b) is *there* in $\mathbb{S}_2$, being multiplied with $\mathbf{i}$.[4]

While observer $\mathbb{O}_1$ remains in $\mathbb{S}_1$ and views z_1 as usual, observer $\mathbb{O}_2$ decides to follow b in $\mathbb{S}_2$. Mathematically, by moving to $\mathbb{S}_2$, observer $\mathbb{O}_2$'s perspective of z changes, being also multiplied by $\mathbf{i}$:

$$z_2 = z_1 \cdot \mathbf{i} = a\mathbf{i} - b$$

So, a becomes $a\mathbf{i}$ and is now imaginary for $\mathbb{O}_2$ because it remained in $\mathbb{S}_1$, which is $\mathbb{O}_2$'s new *there*, while b is no

4 Again, the expression $z = a + b\mathbf{i}$ is the known definition of a complex number.

longer imaginary, but now is $-b$ and O_2 sees it as real because they are both *here* in space S_2.

When observer O_2 decides to return to space S_1, she needs to reverse the move, so her new view of z becomes multiplied with $-\mathbf{i}$:

$$z_3 = z_2 \cdot (-\mathbf{i}) = -z_1\mathbf{i}^2 = -a\mathbf{i}^2 + b\mathbf{i} = a + b\mathbf{i} = z_1$$

After observer O_2's round trip to space S_2, her view of the complex number z in space S_1 reverts to the one she had before the trip.

This exercise is crucial for understanding complex numbers: Both spaces S_1 and S_2 are equally real but depend on the observer. They exchange their perceived nature as *real* or *imaginary* based on whether one is *here* with the observer or *there* in the other space. As noticed, two observers in different spaces can derive different values from the same complex numbers.

Understanding the true nature of the imaginary unit **i** was challenging. In fact, the imaginary unit **i** is more than an operator that rotates a number in the Argand diagram, either counterclockwise ($+\mathbf{i}$) or clockwise ($-\mathbf{i}$) by a 90° angle [114].

9.12 THE UNIT I AS AN INTERSPATIAL OPERATOR

Thus, $\sqrt{-1}$, the imaginary unit $\mathbf{i}$, is the first *interspatial operator*, connecting and switching numbers and observers between two independent spaces.[5]

With defined algebraic operations, this approach reduces complex numbers' properties to those of reals. The square root of -1 becomes more than just a mathematical trick. Bombelli's *wild thought* envisioned a multispace reality.

As Hamilton stated, complex numbers can be defined as ordered pairs of real numbers, with each set on its own space, even without using square roots of -1 [87].

Moreover, complex numbers are also represented as vector spaces, isomorphic to the vector spaces of ordered pairs of real numbers.

It is important to remember that while observer $\mathbb{O}_1$ remains in $\mathbb{S}_1$ and observer $\mathbb{O}_2$ is in $\mathbb{S}_2$, they both view the same numbers a and b as belonging to the real numbers $\mathbb{R}$, even if one number is *here* and the other is *there*.

As in real estate, the meaning of these numbers in multispace physics is a matter of location: New York vs Melbourne.

Explaining that imaginary numbers and 1D spaces are real was the easy part. But this does not explain why

5 The concept of viewing complex numbers as *algebraic couples* of real numbers is not new. In 1835, Irish mathematician William Rowan Hamilton proposed treating complex numbers as ordered pairs of real numbers (a,b) [54].

quantum spaces have incomprehensible properties. To understand why and how these *imaginary spaces* are orthogonal to the 3D space of the universe, we need to learn about Minkowski's story, the biggest mishap in physics.

Thus, we understand Penrose's paradox of describing reality:

> *"Imaginary numbers and spaces are in fact real, but we just did not understand them correctly."*

Now, we understand that reality is a multispace and that all matter in the universe is real but exists in spaces we cannot access. We need to learn about the physical phenomenon of dimensionless and timeless orthogonal spaces that Minkowski probably discovered.

* * *

10 ORTHOGONAL SPACES

10.1 ORTHOGONAL SPACES ARE REAL SPACES

We concluded Chapter 6, Multispace Algebra, by explaining that when a reference frame moves at the speed of light ($v = c$), the real part of the angle δ between two planes becomes $90°$. Thus, the spacetime plane of the observed reference frame is perpendicular to the observer's reference frame, meaning they are *orthogonal*.

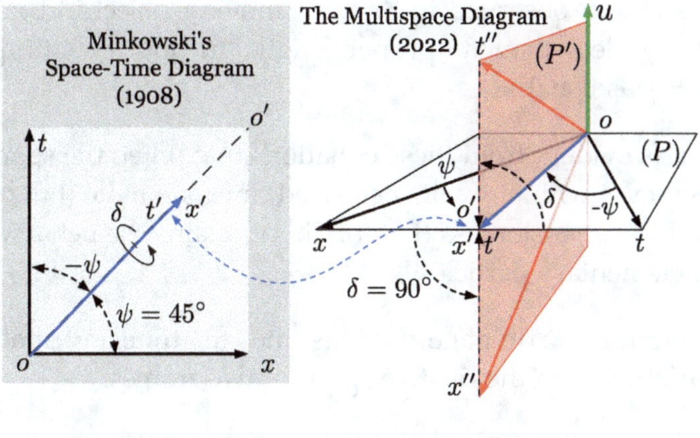

Figure 10.1: Orthogonal spaces are *"real"* spaces

In this scenario, both space and time axes of S_2 project onto the bisector $\overline{oo'}$ of S_1. To comprehend this, consider two core predictions of special relativity derived from the

Lorentz transformations: time dilation (10.1) and space contraction (10.2) [74].

In these equations, **t** and **l** represent measured duration and distance, respectively—not single coordinates. The primed variables denote proper time and proper length, measured in S_2. These transformations map between distinct spaces rather than representing frame swaps or rotations within a unified spacetime metric; boosts are defined as space-to-space mappings.

$$\mathbf{t} = \frac{\mathbf{t}'}{\sqrt{1-\beta^2}} \tag{10.1}$$

Moving clocks in S_2 record less time than an observer in S_1 observes passing, and

$$\mathbf{l} = \mathbf{l}'\sqrt{1-\beta^2}, \tag{10.2}$$

the length observed in S_1 for a moving object in S_2 is always less than its proper length. In both equations, $\beta = \frac{v}{c}$ as usual.

It is evident from these equations that when the space S_2 moves with $v = c$, its perceived time is infinite ($\mathbf{t} = \infty$) and is dimensionless ($\mathbf{l} = 0$). This explains the behavior of elementary particles.

Figure 10.2 on page 105 illustrates the time and space variation with the relative speed of two frames.

The results, combined with the rotation of the spacetime plane described earlier, lead to an important assertion:

Remark 10.1.1. *When the relative speed between two reference frames is the speed of light ($v = c$), the spacetime plane of the observed frame S_2 is perpendicular to the spacetime plane of the*

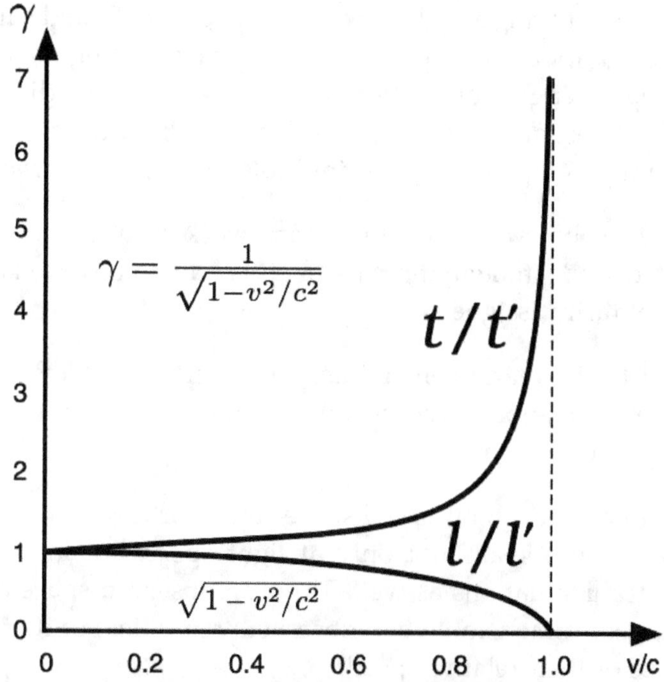

Figure 10.2: Variation of time and length with the relative speed of two spaces

observer's frame S_1. *Regardless of its real properties, any object in the observed reference frame will be dimensionless (in the direction of relative motion) and eternal.* ◇

This assertion may seem debatable. While an empty reference frame can theoretically reach the speed of light, massive objects cannot. Special relativity relies on measurements by physical devices concerning the relative distance and time between objects. Length contraction applies only to lengths measured in the direction of motion;

the other two dimensions remain unaffected. Thus, a 3D object traveling at light speed would appear 2D, and a line would appear as a point only if it travels along its axis. Due to length contraction to zero, the other two dimensions become invisible, rendering any object traveling at light speed point-like and invisible to outside observers.

The observer's spacetime framework is always considered *at rest*, though the roles of observer and observed are arbitrarily assigned.

Despite many constraints, our assertion holds significance. The observed speed v is just one aspect of the extradimension $\mathbf{u}$.

In nature, an observed space can be orthogonal to the observer's space not only at light speed but in other natural situations as well. Thus, an observed space can appear dimensionless to observers regardless of its real size or their relative speed.

Unlike visible spaces, which exist physically within the universe, orthogonal spaces, while embedded in the universe, are simultaneously external to it.

Their curious nature has been acknowledged for decades by both quantum mechanics and string theory. Developed within the single-space paradigm, these theories lack the tools to adequately explain the physical aspects and properties of these seemingly dimensionless pieces of matter.

10.2 DUAL CONCEPT: *HERE* AND *THERE*

In chapter *Physical Nature of Imaginaries*, on page 85, we introduce the dual mathematical concepts of *here* and *there* explained with the dual-space, a theoretical assembly of two 1D orthogonal spaces.

Observers viewing a particle from *here*, i.e., from our universe—*from the particle's outside*—would see a mathematical point. This applies not only to elementary particles but also to separate spaces of any size, including other universes visible only through holes in this universe. Read more about them in chapter 14, *Invisible Universes*, on page 129.

The issue: it is impossible to determine the particle's size, or the size of other universes, while observing *from the outside*. Moreover, it is currently impossible for a real observer to inhabit the space of a particle or visit other universes.

However, theoretical observers *from the inside* of the particle—from *there*—would perceive a physical space with finite, non-zero dimensions.

Remark 10.2.1. From a physical perspective, *everything*, regardless of size or complexity, could exist—unseen yet interactive through its field—in a physical space orthogonal to this universe.◇

10.3 ELEMENTARY PARTICLES AND QUANTUM SPACES

The best examples of spaces orthogonal to our universe are elementary particles, the fundamental constituents of matter. In our space, they do not exist physically; they are dimensionless and timeless (eternal) mathematical points with spatially infinite fields. It is exclusively through their fields that particles interact with one another and with the rest of the universe.

Elementary particles are understood to exist in quantum spaces, a concept rooted in quantum field theory and particle physics.

These particles do not exist in our physical universe as we perceive it; they appear as point-like, dimensionless entities defined by their quantum nature. In this framework, particles are represented by complex numbers and seen as excitations in their respective quantum fields.

Quantum spaces, where elementary particles reside, are orthogonal to the universe's space and invisible from the outside.

These quantum spaces are real and distinct, not merely abstract mathematical constructs. They are embedded within the universe's space but differ from it.

If theoretical observers could view a particle "from the inside," they would perceive a physical space with finite dimensions. This perspective suggests that "absolutely everything" could exist in these external spaces, regardless of size or complexity.

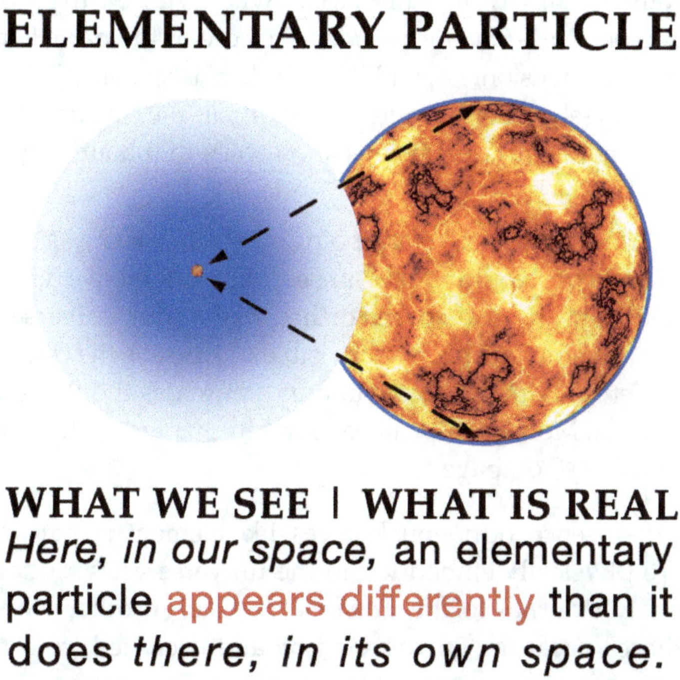

Figure 10.3: Elementary particle.
Credit: CSC CC BY-NC-ND 4.0.

This explanation alters our perception of elementary particles as mere point-like entities that are distinct from our physical universe. These particles are "truly" real and exist in their own universes. We lack knowledge of the actual size of their spaces because our view is confined to the boundaries of our space (our universe.)

This concept also applies to other universes that are invisible to us yet real. It broadens our understanding of the cosmos and highlights a complex interplay between the known and unknown in the cosmic landscape.

Remark 10.3.1. To summarize, when viewed from our perspective—*from the outside*—the core of the particle appears dimensionless, and its boundary is spatially infinite. Conversely, when viewed *from the inside,* the core has finite (and non-zero) dimensions and a very thin boundary.◇

From a physical standpoint, anything of any size or complexity could exist—unseen yet interactive via its field—in physical spaces orthogonal to our universe. A good example of such an invisible universe is the celestial structure NGC 6302, commonly known as the Butterfly Nebula. More details can be found in chapter 14 "Invisible universes" on page 129.

In essence, orthogonal spaces, like elementary particles, are physically embedded in the universe yet simultaneously external to the universe's space. In other words, these spaces are surrounded by and excluded from the universe's spacetime.

* * *

11 QUASI-ORTHOGONAL SPACES

Two unexplained phenomena, *wave-particle duality*, also known as *the photon's dual nature*, and the constancy of *the speed of light*, have perplexed physicists for decades. Both are fundamental to understanding the basic structure of reality, without which our knowledge remains uncertain.

11.1 WAVE-PARTICLE DUALITY

According to Encyclopaedia Britannica:

> *"Wave-particle duality is a cornerstone of quantum mechanics, encapsulating the idea that particles such as electrons and photons exhibit both wave-like and particle-like properties [41]."*

This initially puzzling concept led to significant developments in physics. Interpretations like the Copenhagen interpretation, Many-Worlds interpretation, and pilot-wave theory aim to provide clearer conceptual understanding. Essentially, every particle or quantum entity may be described as either a particle or a wave, highlighting the inadequacy of classical concepts *particle* or *wave* to fully describe quantum-scale objects.

As Albert Einstein wrote [5]:

> *It seems as though we must use sometimes the one theory and sometimes the other, while at times we may use either. We are faced with a new kind of difficulty. We have two contradictory pictures of reality; separately neither of them fully explains the phenomena of light, but together they do.*

In 1921, Einstein received the Nobel Prize in Physics for his discovery of the law of the photoelectric effect, i.e., that light is quantized in photons, but he could not explain why.

> *"Through the work of Max Planck, Albert Einstein, Louis de Broglie, Arthur Compton, Niels Bohr, Erwin Schrödinger, and many others, current scientific theory holds that all particles exhibit a wave nature and vice versa [51]. "*

> *"Although the use of the wave–particle duality has worked well in physics, the meaning or interpretation has not been satisfactorily resolved; see interpretations of quantum mechanics.*
> *Also, Bohr regarded the* duality paradox *as a fundamental or metaphysical fact of nature [51]."*

11.2 SPEED OF LIGHT

The constancy of the speed of light in a vacuum, established by Albert Einstein's theory of relativity, is fundamental to modern physics. It states that the speed of light is constant for all observers, regardless of their motion.

> *"The speed of light in vacuum, commonly denoted c, is a universal physical constant important in many areas of physics. The speed of light c is exactly 299,792,458 metres per second (approximately 300,000 kilometres per second; 186,000 miles per second; 671 million miles per hour). According to the special theory of relativity, c is the upper limit for the speed at which conventional matter or energy, and thus any signal carrying information, can travel through space [43]."*

All electromagnetic radiation, including visible light, travels at the speed of light. For many practical purposes, light and other electromagnetic waves seem to propagate instantaneously. However, over long distances and in sensitive measurements, their finite speed becomes noticeable.

To deepen the understanding of wave-particle duality and the constancy of the speed of light, we introduce two concepts: *Quasi-orthogonal Spaces* (QOS) and the *Most Fundamental Cruise Control* (MFCC) of nature.

11.3 QUASI-ORTHOGONAL SPACES

The Quasi-orthogonal Spaces concept suggests that quantum entities, like photons, exist in spaces that transition between connected and disconnected (orthogonal) states relative to the universe's space.

11.4 MOST FUNDAMENTAL CRUISE CONTROL

The Most Fundamental Cruise Control (MFCC) of nature details the phases of these transitions.

Minkowski could not have imagined the two orthogonal spaces P and P' in his drawings in Figure 3.1 on page 21 unless he had first discovered his 2D diagram introduced in the Cologne lecture [6].

In Figure 10.1 on page 103, we present a 2D diagram alongside a 3D Multispace Diagram, displaying orthogonal planes P and P'.

To grasp these concepts, consider two observers, $\mathfrak{O}_1$ and $\mathfrak{O}_2$, in orthogonal spaces $\mathfrak{S}_1$ (or P) and $\mathfrak{S}_2$ (or P'). Equation 6.17 on page 53 describes the position of observer $\mathfrak{O}_2$ in her space $\mathfrak{S}_2$, where speed $v = c$ is real and *here* for her, but imaginary and *there* for observer $\mathfrak{O}_1$ in $\mathfrak{S}_1$. When spaces are orthogonal, their projections on each other are zero, preventing mutual visibility. This is the case for elementary particles in our universe—quantum spaces [6]—and of universes surrounding our own [7].

In physics, photons are considered energy quanta—they are the smallest discrete packets of energy associated with electromagnetic radiation such as light. In quantum theory, energy exchanges between matter and electromagnetic fields occur in these quantized units, where each photon carries a specific amount of energy given by $E = h\nu$, with h being Planck's constant and ν the frequency of the photon.

This means that light and all electromagnetic radiation do not transfer energy continuously, but in multiples of this fundamental unit—the photon, or quantum of energy.

In multispace, photons exist in their own spaces that travel with speed v.

Consider a photon leaving its source. Initially, the photon is in an energy state in its space $\mathbb{S}_2$ (or P'), accelerating its speed v toward c within the universe's space $\mathbb{S}_1$ (or P). As long as $v < c$, the angle of $\mathbb{S}_2$ is $\delta(v) = 90° - \epsilon$, where $\epsilon > 0°$, meaning space $\mathbb{S}_2$ has a projection on space $\mathbb{S}_1$. The two spaces are connected and visible to each other. The energy in $\mathbb{S}_2$ travels through $\mathbb{S}_1$ as a wave.

When $v = c$, the angle $\delta(v) = 90°$, making the two spaces orthogonal and disconnected. At this point, energy is confined into a particle, transforming into a photon with mass that begins to decelerate.

As the particle decelerates ($v < c$), $\delta(v) < 90°$, and space $\mathbb{S}_2$ reconnects with space $\mathbb{S}_1$, becoming visible again as a wave.

This cycle of alternating between wave and particle states repeats indefinitely (or until the photon interacts with an object), continuously flipping between states. Simply put, photons are quasi-orthogonal spaces, and this process is the Most Fundamental Cruise Control (MFCC).

The MFCC addresses several open questions in theoretical physics, including wave-particle duality, the constancy of the speed of light, and the foundation of Maxwell's equations. It also accounts for quantum entanglement, where seemingly separate objects belong to the same independent space, yet appear distant in our universe, challenging classical intuitions about separability [6].

11.5 ON ELECTROMAGNETIC TRAVEL

A significant application of the MFCC is electromagnetic travel near the speed of light. For example, an interplanetary vessel traveling at 1% of the speed of light would take 5 hours, 3 minutes, and 20 seconds to reach Mars at its closest distance of about 54.6 million kilometers (33.9 million miles).

Electromagnetic travel requires the development of two technologies: (1) separating the vessel's space from the universe's space, and (2) propelling the vessel to "ride" on the universe's space.

11.6 MAXWELL'S EQUATIONS

Between 1861 and 1865, James C. Maxwell published several papers, laying the foundation of electromagnetism[31], later known as *Maxwell's equations* [69].

These equations are a set of mathematical formulae that, together with the Lorentz force law, form the basis of classical electromagnetism, classical optics, and electric circuits. They describe how electric and magnetic fields are generated by charges, currents, and changes in the fields.

According to the theory of relativity, electric and magnetic fields are components of a single electromagnetic field.

Maxwell first used the equations to propose that light is an electromagnetic phenomenon. In an 1862 paper, he derived the speed of light *c* from the expression of the velocity of the electromagnetic wave in relation to the vacuum constants.

Today, Maxwell's equations are crucial but have a significant drawback: they show *how* electromagnetic fields propagate but do not explain *why* they move this way. However, this phenomenon is perfectly explained by orthogonal spaces and the Most Fundamental Cruise Control of nature.

* * *

Part V

ULTIMATE SECRETS OF COSMOLOGY

12 THEORY OF DECAYING UNIVERSES

What is the extent of physical reality? We're confident that it's more extensive than the domain that astronomers can ever observe, even in principle. That domain is definitely finite. That's essentially because, like on the ocean, there's a horizon that we can't see beyond. And just as we don't think the ocean stops just beyond our horizon, we expect galaxies beyond the limit of our observable universe.

—Lord Martin Rees [92]

12.1 REALITY IS A NESTED OMNISPACE

The Theory of Decaying Universes suggests that reality is a vast, nested structure called an *omnispace,* encompassing countless universes interacting through energy and matter leakage via *cosmholes.* This framework implies that universes generate subsequent generations, with each transition producing more complex forms of matter. Galaxies, clusters, and superclusters in our universe are boundary effects of this process, linking our observable reality to a chain of preceding universes.

In recent decades, the multiverse concept has gained traction among scientists. In 1998, British cosmologist and astrophysicist Lord Martin Rees suggested that *our entire universe may be just one element—one atom, as it were—in an infinite ensemble: a cosmic archipelago [93].*

While Rees's vision of a multiverse is compelling, the *Theory of Decaying Universes* extends this idea further, positing that reality is a nested omnispace—a comprehensive space containing myriad universes and serving as the ultimate container for all conceivable forms of existence.

The term *omnispace* may seem speculative, as it has not yet been integrated into mainstream cosmological research. Unlike traditional terms *multiverse* or *omniverse,* which focus on collections of universes, omnispace transcends these limitations. It envisions reality as a holder or space—an *omnispace*—offering a more comprehensive idea and serving as the ultimate container not only of universes but also of any conceivable or inconceivable structures of existence.

This concept transcends the limitations of our current language and conceptual frameworks, which may overlook other fundamental forms of existence by focusing too narrowly on *multi- or omni- verses.* By embracing the notion of *omnispace,* we challenge restrictive categorical thinking, broadening our understanding to potentially encompass myriad forms of existence beyond what traditional definitions permit.

As we have explained in section 1.3 "The omnispace" on page 6 of chapter 1 "Why a New Paradigm" on page 3, the omnispace is a dynamic, multilayered structure, analogous to a set of Matryoshka dolls, where the omnispace contains numerous layered spaces. This nested configuration implies that our universe, other universes, and even elementary particles are equivalent spacetime frameworks governed by the same physical laws.

As discussed in chapter 10 on page 103, the theory is based on two foundational principles: reality consists of numerous independent spaces, and many of these spaces are relativistically orthogonal to our universe's spacetime or to each other.

12.2 OMNISPACE STRUCTURE

The Theory of Decaying Universes suggests that reality started with an initial omnispace G_0, consisting of pure energy. This energy escapes through *cosmholes*—vortices similar to wormholes linking universes—into successive universes, creating a hierarchical chain. (The detailed mechanics of cosmholes are discussed in chapter 15, *The Cosmhole Hypothesis*, on page 133.)

The structure begins with the initial omnispace G_0 releasing pure energy into the first generation of universes, G_1. This process converts some energy into elementary particles, while some energy passes through. The G_1 universes leak energy into G_2 universes. Some of this energy transforms into ionized hydrogen gas due to boundary effects, creating structures like the Orion Nebula in our universe.

The G_2 universes leak into G_3 universes, converting hydrogen into helium and other elements. Some material remains unchanged, akin to the Butterfly Nebula's composition, rich in hydrogen and helium. Successive generations, G_n, continue this process. Each transition produces more complex matter at universe boundaries.

Boundary effects contribute to galaxy formation and larger cosmic structures. The theory proposes that free elementary particles—quarks, leptons, and gauge bosons—persist from earlier universes into ours. Their presence supports the hypothesis, yet specific examples or anomalies are not detailed.

12.3 OUR UNIVERSE IN THE CHAIN

We do not know our universe's exact position in the sequence of decaying universes, so we designate it as the reference point U_0. Preceding universes are labeled U_{-1}, U_{-2}, etc. The theory suggests that structures within U_0—such as galaxies, clusters, and superclusters—are boundary effects from earlier universes. The Milky Way Galaxy exemplifies a boundary effect of U_{-1} leaking into U_0. It contains stars, planets, and complex matter, reflecting cumulative evolution through previous generations.

The Local Group, which includes the Milky Way, is connected to leakage from U_{-2} to U_{-1}, affecting U_0. The Virgo Supercluster, containing the Local Group, arises as a boundary effect from U_{-3}. The Laniakea Supercluster, a larger structure that includes the Virgo Supercluster, is linked to leakage from U_{-4}. The name "Laniakea" is Hawaiian for "immense heaven," reflecting its vast scale and Hawaiian discovery heritage.

This nested hierarchy suggests that each level of cosmic structure in our universe corresponds to interactions with previous universes. However, the precise mechanism connecting these scales needs further clarification.

12.4 IMPLICATIONS OF THE THEORY

The Theory of Decaying Universes offers a new framework for understanding the origin and complexity of matter and cosmic structures. It considers galaxies and clusters as boundary effects of inter-universe leakages, connecting local phenomena like nebulae and stars to a vast, unseen reality. The survival of elementary particles from other universes provides a testable hypothesis, potentially observable through anomalies in particle physics or cosmology.

The Theory of Decaying Universes redefines traditional cosmological models by viewing reality as a nested omnispace. It suggests a dynamic, hierarchical structure where our universe is one thread in a complex tapestry of decaying universes, linked through cosmholes as boundary effects. This perspective expands our understanding of existence, prompting further exploration into reality and its hidden dimensions.

* * *

13 | INTENTIONALLY LEFT BLANK

This chapter is intentionally left blank.

* * *

14 INVISIBLE UNIVERSES

The concept of *invisible universes* builds on the foundation established in the previous section on elementary particles and quantum spaces, as detailed in section 10.3 on page 108.

These invisible universes are theorized to leak energy and matter into our universe, yet remain hidden from our perception. This invisibility arises because our observation is limited to our universe, unless there are openings that connect it with another.

A remarkable example that aids in understanding this phenomenon is the celestial structure NGC 6302, commonly known as the Butterfly Nebula, illustrated in Figure 14.1.

This nebula stands out due to its complex structure, making it one of the most intricate planetary nebulae observed. A mysterious dark lane, highlighted by a red oval in the image, runs through the nebula's central region, obscuring the central star across all wavelengths, as noted by Matsuura et al. [3].

The spectrum of NGC 6302 reveals that this central star is among the hottest known, with a surface temperature exceeding $250,000°C$.

Detecting this star is challenging due to its extreme heat, the surrounding dusty torus, and the bright back-

Figure 14.1: NGC 6302, the Butterfly Nebula. Credit: NASA, ESA, Hubble CC BY-SA 3.0 IGO.

ground of the nebula. Traditionally, it is assumed that the star responsible for forming this nebula must have been exceptionally large.

The precise symmetry of the Butterfly Nebula's wings suggests an alternative interpretation. Instead of originating from a single massive star, the narrow black region at the nebula's center might belong to a vast, hidden universe beyond our observation. This hidden universe could project its gaseous wings—composed of immense quantities of oxygen (blue), hydrogen (green), and nitrogen (red)—into our universe, creating a structure spanning over three light years.

The figure below visually represents the universe that may be hidden from our view:

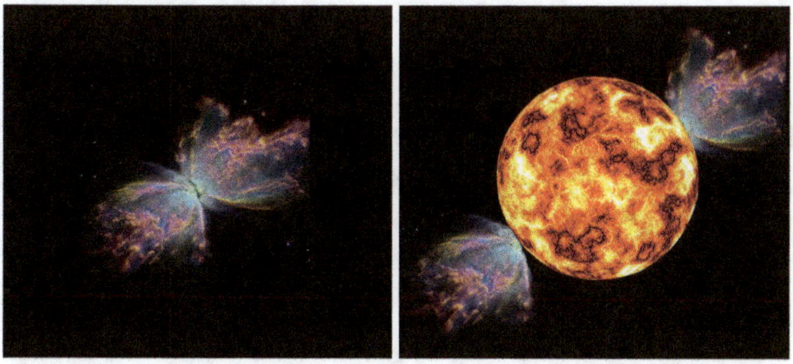

Figure 14.2: The Butterfly Nebula: What we see and what is real but invisible. Credit: CSC CC BY-NC-ND 4.0.

* * *

15 | COSMHOLE HYPOTHESIS

The Cosmhole Hypothesis suggests a complex mechanism for the origin and evolution of celestial bodies, such as stars, planets, and universes. At its core is the *Cosmhole Effect*, where energy and matter from one universe enter another, like ours, through vortices.

In our universe, this transfer creates the cosmic objects we observe. The process is cyclical, with matter evolving into more complex forms each time it crosses a universe's boundary.

For instance, pure energy from the *omnispace*—the original universe—or another universe partially converts into elementary particles, such as electrons and neutrinos, upon entering our universe. This aligns with established physics, as stars like the Sun produce vast quantities of these particles.

The exact workings of the Cosmhole Effect remain unclear, but the hypothesis draws inspiration from René Descartes' vortex theory. Descartes observed that celestial bodies exhibit rotational motion, simultaneously swirling around themselves and other centers. This is evident in Earth's orbit around the Sun, the Sun in the Milky Way, and the broader movement of the Milky Way within the universe.

Building on this idea, the Cosmhole Hypothesis suggests that vortexes create channels—termed *cosmholes*—

through which energy and matter leak between universes. Unlike wormholes, which connect distant points within the same universe, cosmholes link different universes. A cosmhole has two ends (see Figure 15.1). The first end opens into the source universe, appearing as a *Blackhole* that absorbs energy and matter. The other end opens into a receiving universe, like ours, where it functions as a *white hole,* releasing the absorbed material. We named this transfer the *Cosmhole Effect*. It is the mechanism by which matter and energy enter all universes, including ours. Since the cosmhole exists outside both universes, it remains invisible from within them. In our universe, white holes appear to emerge from nowhere, expelling matter through various celestial bodies.

Figure 15.1: Vortex. Credit: Scientific American

One source universe, U_{-1}, can be envisioned as a space filled with "meta-magma," a scorching, heterogeneous fluid of energy and condensed matter, with temperatures possibly exceeding $100,000,000°K$. This meta-magma is in constant motion, forming massive streams of matter and energy. Occasionally, these streams collide or merge, generating powerful spinning vortexes.

Some vortexes grow so intense that they carve worm-like channels, or cosmholes, piercing the source universe's boundary and draining meta-magma into other universes. As it travels through the cosmhole, the meta-magma may decay into *magma,* like the molten material found in

Earth's interior. This magma becomes the building block of all matter in our universe, from the fuel of stars to the inorganic substances on planets, and even the cosmic dust.

Depending on the meta-magma's composition in the source universe's region where a cosmhole forms, two types of white holes can emerge in its receiving universe. *Confined white holes,* or *C-white holes,* are embedded within celestial bodies, like the Sun's core or Earth's inner core, remaining hidden. In contrast, *open white holes,* or *O-white holes,* manifest visibly in the cosmos as quasars or nebulae, such as the M42 Nebula in Orion or the Planetary Nebula NGC3132.

The Cosmhole Hypothesis describes the expansion of celestial bodies, like Earth, through a five-phase process. Initially, the body experiences rapid growth. Over time, this expansion slows as the growth rate decreases. Eventually, the expansion halts temporarily. Later, a second expansion may occur, followed by accelerated growth where the rate increases again. This phased growth happens when the white hole's magma pressure is low enough for the body's solid crust to resist further expansion. For example, if the pressure were even lower, Earth might have stabilized at a diameter of just 50 kilometers. This behavior defines a *Low Pressure Confined white hole,* or LPC-white hole.

When the magma flow is strong enough to sustain the expanding body's heat, growth continues unimpeded, and the celestial object expands rapidly. This process describes the formation of stars like the Sun and is typical of a *High Pressure Confined white hole* (HPC-white hole). The distinction between LPC-white holes and HPC-white holes originates in the source universe, where the vortex size determines the outcome. A massive vortex can create a black hole leading to an HPC-white hole, forming a star like the Sun. Smaller, local vortexes on the sides of this central vortex produce LPC-white holes, resulting in planets. For instance, Earth and the Moon, which orbit together, likely originated from the same local vortex, each developing its own LPC-white hole through a process not yet fully understood.

According to this hypothesis, the creation of our universe is an ongoing process that began approximately twenty billion years ago and will continue indefinitely. New cosmholes can form at any time, creating stars, planets, or nebulae. This perspective explains the ongoing and accelerating expansion of our universe. It challenges traditional cosmic origin theories, suggesting that stars shine not due to thermonuclear reactions but because they are formed by matter and energy entering through cosmholes.

A star's brightness is influenced by the size of its cosmhole. A large cosmhole ensures a steady flow of metamagma, keeping the star bright for extended periods. Conversely, a smaller cosmhole may lack sufficient material, causing the star to cool and dim, eventually fading as a dying star.

The hypothesis redefines black holes. Traditionally seen as regions from which nothing escapes, black holes are now viewed as points where energy and matter transfer between universes. For instance, the supermassive black hole Sagittarius A* at the Milky Way's center is considered a conduit to another universe, evidenced by the galaxy's rotational motion around it. The Milky Way, part of the Local Group—a cluster of galaxies including

Andromeda (M31) and Triangulum (M33)—illustrates the interconnectedness of universes via cosmholes.

The Cosmhole Hypothesis provides a framework for understanding the universe's origin, structure, and evolution. It suggests our universe is at the receiving end of other universe through cosmholes. However, the exact number of source universes is unknown. This hypothesis offers new insights into celestial bodies, cosmic expansion, and the processes driving cosmic evolution.

* * *

16 TDU VS CURRENT SCIENCE

It is essential to compare the new interpretation provided by the Theory of Decaying Universes (TDU) with centuries of accumulated knowledge about the cosmos, galaxies and nebulae.

Understanding stellar populations is crucial in astrophysics, offering insights into star formation processes and the evolutionary history of galaxies. A star's metal content reveals much about its formation environment. (Note that in astronomy, *metals* refer to elements other than hydrogen and helium.)

The key question is whether our hypothesis is validated by stellar populations or dismissed as pseudoscientific.

In astronomy, stars are classified into three populations:

- Population III stars (Pop III):

Age: Hypothetical and very ancient.

Metallicity: Assumed to contain almost no metals. These stars would have formed from the primordial material of the Big Bang, consisting mainly of hydrogen, helium, and traces of lithium.

Location: Not directly observed, but thought to have existed in the early universe and may have reionized the cosmos.

TDU view: These stars have not been directly observed because they never existed. They lack sufficient metallic material to become bright stars. Nebulae, common in galaxies, provide the necessary conditions for star formation. However, not all galaxies contain nebulae. Current astronomical surveys estimate several thousand nebulae in the Milky Way, but their exact number remains unknown.

- Population II stars (Pop II):

Age: Older stars.

Metallicity: Lower metallicity, indicating formation earlier in the universe's history, before many heavier elements formed in stars.

Location: Typically found in the bulges and halos of galaxies and in globular clusters.

TDU view: This stellar population indicates that older stars formed during earlier generations of universes, marking the start of the chain of decaying universes.

- Population I stars (Pop I):

Age: These stars are relatively young.

Metallicity: They possess higher metallicity.

Location: They are typically found in spiral arms, young clusters, and galaxy disks.

Examples: The Sun is a Population I star.

TDU view: Young stars have evolved from subsequent universe generations.

16.1 DIFFERENCES BETWEEN TDU AND SCIENCE

- The universe is expanding.

  This concept is fundamental to modern cosmology, first observed by Edwin Hubble in the late 1920s. He noted that galaxies are moving away from us, with more distant galaxies receding faster than closer ones (Redshift of Distant Galaxies).

TDU view: The formation of all universes, including ours, is ongoing, causing expansion. Many universes from various generations leak into ours through boundary effects. Our universe and others are constantly expanding due to the influx of matter.

On their side, these are black holes; on our side, they appear as white holes: galaxies, stars, planets, and other celestial bodies.

TDU view: All black holes in our universe leak into other universes, where they appear as white holes.

- Current challenges and unsolved mysteries

  - The mystery of Dark Matter: Its effects are observed, yet its true nature remains unknown.

- Accelerating universe: Dark energy's role in the universe's expansion acceleration.

- Key points:

  Scientists estimate that dark matter and dark energy make about 95% of the universe. Only about 5% consists of the "ordinary" matter we know, such as stars, planets, galaxies, and other celestial objects.

  These concepts challenge our current understanding of physics.

  - Dark matter is considered a form of matter that neither emits, absorbs, nor reflects light. It does not interact with electromagnetic forces, making it invisible. However, its existence is inferred from its gravitational effects on visible matter, radiation, and the universe's large-scale structure. Unlike ordinary matter, which comprises protons, neutrons, and electrons, dark matter is hypothesized to consist of undiscovered subatomic particles.

TDU view: Dark matter is unnecessary for explanation. The hypothesis that matter consists of undiscovered subatomic particles is meaningless.

Dark matter does not exist!

- Dark energy is a mysterious force driving the universe's accelerated expansion, seemingly causing a repulsive effect that increases the expansion rate.

TDU view: Dark energy is unnecessary to explain the universe's expansion. The universe expands by receiving energy and matter from white holes opening from other universes: stars, nebulae, planets, etc. This hypothesis of a "mysterious force with a repulsive effect" is also meaningless.

Dark energy does not exist either!

16.2 DNA OF STARS

The composition of stars provides insights into their origins and journeys, shedding light on the Milky Way's history. With Data Release 3, Gaia offers the largest chemical map of the galaxy, combined with 3D motions, extending from our solar neighborhood to nearby smaller galaxies.

A star's chemical composition is like its DNA, offering crucial information about its origin.

Gaia's data suggests that some stars in our galaxy are composed of primordial material, as only light elements (hydrogen and helium) were formed during the Big Bang.

- According to TDU, the primordial material from universes G_1 and G_2 consists of hydrogen and helium, confirming the Big Bang nucleosynthesis.

Heavier elements, referred to as "metals" by astronomers, are formed inside stars like our Sun. These stars contain more "heavy metals" as they are composed of matter enriched by previous generations of decaying universes.

- According to TDU, heavier elements are formed by successive boundary effects during transitions to newer decaying universes.

* * *

Part VI

COSMIC SCALE VALIDATION

17 COSMIC SCALE

Our entire universe may be just one element—one atom, as it were—in an infinite ensemble: a cosmic archipelago.
— Professor Martin Rees [35]
British cosmologist and astrophysicist

17.1 INTRODUCTION

The terms "reality" and "universe" are often used interchangeably in casual conversation, but in precise contexts, they have distinct meanings.

The term "universe" typically refers to the totality of space, time, matter, and energy. When scientists mention the universe, they mean the entirety of physical existence, including all galaxies, stars, planets, subatomic particles, forces, and everything in between.

"Reality" is a broad philosophical term that encompasses everything real or existing, whether perceived, conceptualized, or beyond our understanding. It includes physical universes, ideas, thoughts, feelings, dreams, mathematical truths, and more. The question of reality's nature prompts exploration of what is truly "real' and how we can know it.

A key aspect of our argument involves reconsidering the Milky Way galaxy's structure. This prompts a reassessment of our cosmic origins and the universe's overall structure.

To advance this paradigm shift, we examine certain nebulae and galaxies, hypothesizing they may not be conventional celestial bodies. This theory challenges established ideas about the Big Bang, dark matter, and dark energy, reimagining matter formation through a multistep mechanism.

Data from GAIA, combined with the Theory of Decaying Universes, could significantly alter our understanding of physical reality, encouraging researchers to explore beyond the observable universe. They propose that understanding our universe requires estimating the dimensions and shape of the universe from which the Milky Way emerged.

17.2 UNVEILING THE MILKY WAY MYSTERIES

17.2.1 Insights From Gaia Mission

Our Solar System is part of the Milky Way Galaxy, which contains nearly 1.7 billion stars and takes approximately 225-250 million years to orbit its center. Its average velocity is about 828,000 km/h (approximately 514,000 mph). Despite this high speed, the vast distance involved results in an extremely long orbital period.

The GAIA mission, a division of the European Space Agency, is creating an accurate three-dimensional map of nearly two billion celestial objects in our galaxy and beyond. It maps their motion, brightness, temperature, and composition [97]. This comprehensive catalog of stellar data enables researchers to address unresolved questions about the origin, structure, and evolutionary history of our galaxy.

Gaia's extensive catalog of celestial objects enhances our understanding of the solar system, galaxy, and fundamental physics of the universe.

The name "GAIA" was initially an acronym for "Global Astrometric Interferometer for Astrophysics," reflecting the planned use of interferometry on the space probe. Despite changes in methodology, the name Gaia persists for project continuity.

The image in Figure 17.1 above shows an artist's impression of the Milky Way and, above it, an overlay showing the location and density of a sample of young stars from Gaia's Data Release 3 (in yellow-green). The "You are here" sign points to the Sun [99].

However, it is important to note that this image was calculated by Gaia Data Release 3 and is not the image we see from Earth.

The actual image of the Milky Way as seen from Earth is in Figure 17.2 and appears asymmetrical. The lower side of the galaxy seems more elongated than the upper side.

Please note this discrepancy, as there is a hidden reason for it.

Figure 17.1: You are here in Milky Way. Credit: ESA Gaia, DPAC, CC BY-SA 3.0 IGO.

17.2.2 Gaia Data Release 3 (Gaia DR3)

In June 2022, ESA's Gaia mission released "Gaia DR3," a new dataset about the Milky Way [94][95].

Gaia's DR3 offers enhanced details for nearly two billion stars in our galaxy. The catalog provides information on chemical compositions, stellar temperatures, colors, masses, ages, and radial velocities. Much of this data was derived from newly published spectroscopy, which splits starlight into individual colors. The dataset also includes specific subsets of stars, such as those with variable brightness.

Figure 17.2: The Milky Way viewed from the Earth. Credit: ESA Gaia, DPAC, CC BY-SA 3.0 IGO.

Additionally, this dataset features the largest catalog of binary stars to date, thousands of solar system objects like asteroids and planetary moons, and millions of galaxies and quasars beyond the Milky Way.

17.2.3 Apparent Inconsistency of Gaia DR3

Gaia DR3 reveals an inconsistency by portraying the Milky Way galaxy as both asymmetric and symmetric.

On one side, the Milky Way's image in Figure 17.2 on page 150, known for centuries and presented by ESA's Gaia DR3, shows the galaxy as asymmetric [65], with the top compressed and the bottom expanded.

Gaia explains, "more than 30 million Milky Way stars are moving towards or away from us," and "we can clearly see that on average the stars do not rotate with circular motions around the center of the galaxy." This indicates our Milky Way is not symmetrical about its axis.

On the other side, while the above image of the Milky Way as seen from Earth is real, this explanation for the movements of its stars is not accurate.

Thanks to the Gaia DR3 video[96], we can see at second 38 that the Milky Way is perfectly symmetrical about its axis when viewed from above, as in Figure 17.3 below. The Milky Way is not asymmetrical at all.

Figure 17.3: The Milky Way seen from above. Credit: ESA Gaia, DPAC, CC BY-SA 3.0 IGO

Over the years, scientists have noted that the Milky Way displays symmetry in its spiral arms and central bulge when viewed broadly. However, local asymmetries arise due to uneven gas distribution, star formation, and past interactions with other galaxies.

A closer examination of the video reveals this inconsistency. At second 32, the galaxy's plane rotates from the compressed image, which should correspond to our view—as shown in Figure 17.2—until the full image of the planar galaxy is visible from above at second 38.

At second 33, the view differs from Figure 17.2. Instead, we see the inclined plane in Figure 17.4, which maintains the proportions between its top and bottom.

Figure 17.4: The image we should see if the Milky Way were planar, but it is not. Credit: ESA Gaia, DPAC, CC BY-SA 3.0 IGO.

This image is clearly the vertically compressed version of Figure 17.3 and differs significantly from the real image 17.2 seen from Earth.

How can Gaia scientists describe the Milky Way as asymmetrical in one article and symmetrical in the other?

One possibility is that they cannot find a reasonable explanation within the current universe paradigm for why the lower part of the real galaxy is so elongated compared to the upper part.

In fact, since the late 1950s, astronomers have believed the Milky Way galaxy has an intriguing property: it is warped. They suggest its disk is not flat but curves upwards on one side and downwards on the other. The cause of this warping has been debated among astronomers for years.

However, Gaia scientists do not support the hypothesis of the Milky Way's warping. They argue that the galaxy cannot be both asymmetric and symmetrical simultaneously. They propose that the Milky Way's curvature is not a simple deformation but forms a convex structure. This shape cannot be explained by the current paradigm of a unique universe.

17.2.4 Milky Way Galaxy Is Leaked From Another Universe

In a multiverse paradigm, this deformation can occur if the Milky Way's "galactic disk" is not planar or warped at both ends but convex, appearing in our universe as warped around another universe's boundary.

Simply put, the Milky Way is a convex galaxy (a new classification), leaked as a "boundary effect" of an invisible, spherical universe.

The illustration in Figure 17.5 on page 155 is clear. The Milky Way is a convex galaxy, a result of a "boundary

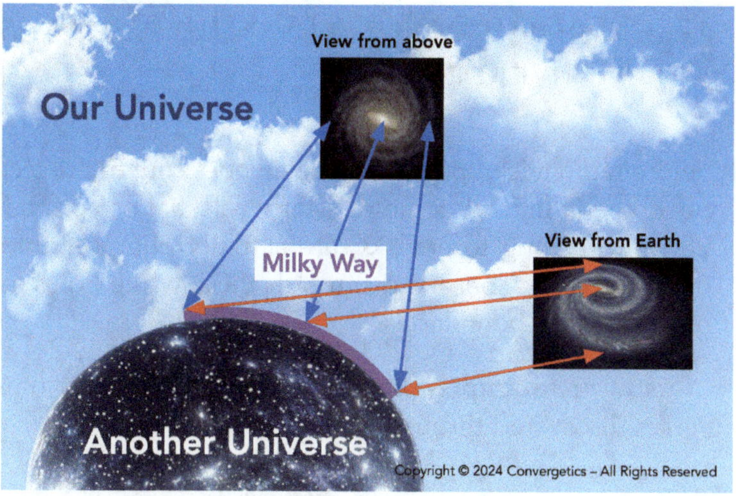

Figure 17.5: Milky Way galaxy is warped on another universe. Credit: CSC CC BY-NC-ND 4.0.

effect" from an invisible, quasi-spherical universe. This represents a new classification for galaxies. Insights into these invisible universes are discussed in Chapter 14, "Invisible universes," on page 129.

Assuming each star in the Milky Way represents a boundary effect between universes, advanced computing could analyze Gaia DR3 data to determine the dimensions and shape of the universe from which our galaxy originates.

17.3 GALAXIES AND STARS IN OTHER UNIVERSES

Before radio telescopes, optical telescopes couldn't detect phenomena like white holes. Many celestial objects, believed to belong to our sky, are actually glimpses into other universes, observed through these white holes at our universe's boundaries.

Below are examples of direct glimpses into other universes discovered through the TDU.

17.3.1 Polar Ring Galaxy (NGC 4632) With a Hidden Polar Ring

The galaxy NGC 4632 is among the first instances when a polar ring was detected via a radio telescope survey. Viewed through an optical telescope, it appears like many other galaxies, as shown in Figure 17.6.

Yet, observed with the highly sensitive CSIRO's ASKAP telescope, alongside optical data from the Subaru telescope, the galaxy is encircled by a ring of cold hydrogen gas orbiting at a 90-degree angle to its spiral disk. See Figure 17.7.

Astronomers employ virtual reality to distinguish hydrogen gas in the galaxy's disk from the ring and track its orbital motion. They propose that polar rings form via gravitational interactions between galaxies or when hydrogen gas from the cosmic web accumulates in a galaxy's ring, prompting star formation.

Figure 17.6: The Polar Ring galaxy viewed with an optical telescope. Credit: Sloan Digital Sky Survey.

Cosmologists suggest this structure could be a polar ring orbiting the galaxy at a 90°angle, providing insights into the galaxy's dark matter distribution and evolutionary theories.

However, TDU asserts that the cosmic ribbon near the Polar Ring nebula is unrelated to the galaxy. It is formed by hydrogen gas escaping through a white hole's boundary between that galaxy's universe and ours. The ribbon's exterior marks our universe's border, while the Polar Ring galaxy exists in the other universe's space. Watch the video here [63].

Figure 17.7: The Polar Ring galaxy has a hidden polar ring seen with the new ASKAP telescope. Credit: J. English, N. Deg & WALLABY Survey, CSIRO/ASKAP, Subaru Telescope.

17.3.2 Orion Nebula (M 42)

The Orion Nebula, depicted in Figure 17.9, is approximately 1300 light-years from Earth in the "Sword" of the constellation Orion the Hunter. The image captures a region measuring 4 by 2.75 light-years. Known as M42, this nebula is among the most extensively studied celestial objects. Data primarily originate from various telescopes and instruments across different wavelengths.

Optical observations from the Hubble Space Telescope revealed intricate details of the nebula's structure, highlighting its bright central region where the Trapezium

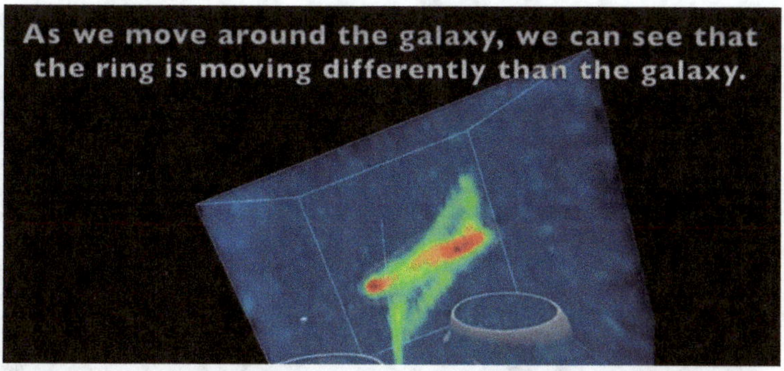

Figure 17.8: The cosmic ribbon marks the boundary of a white hole.Credit: J. English, N. Deg, CSIRO/ASKAP, and Subaru Telescope.

cluster resides. Infrared data from the Spitzer Space Telescope offered insights into the nebula's star-forming regions, hidden by dust. In the radio range, the Very Large Array (VLA) provided a view of the cold gas and molecular cloud dynamics.

Ground-based telescopes, such as the Keck Observatory and the Very Large Telescope (VLT) of the European Southern Observatory (ESO), provided high-resolution spectroscopic data crucial for understanding the nebula's chemical composition and physical processes. Multiwavelength surveys were essential for mapping complex magnetic fields, gas flows, and ionization zones.

This pre-2023 data established a foundation for comprehensive multidisciplinary studies, assisting astrophysicists and cosmologists in forming hypotheses about stellar evolution, planet formation, and interstellar material dynamics.

Figure 17.9: Orion Nebula (M 42): The Hubble View. Credit: NASA, ESA, M. Robberto (STScI/ESA)

In October 2023, the NIRCam instrument on the James Webb Space Telescope (NASA/ESA/CSA) provided a groundbreaking view of the Orion Nebula with its advanced infrared capabilities. (See Figure 17.10.)

The image, captured with exceptional thermal infrared sensitivity, reveals a complex interplay of ionized gas, dust, and molecules. Thousands of new stars have been discovered, ranging from 40 times to less than 0.1 times the mass of the Sun [98]. The Trapezium cluster stars have carved a cavity in the massive molecular cloud from which they formed, clearly visible in the image.

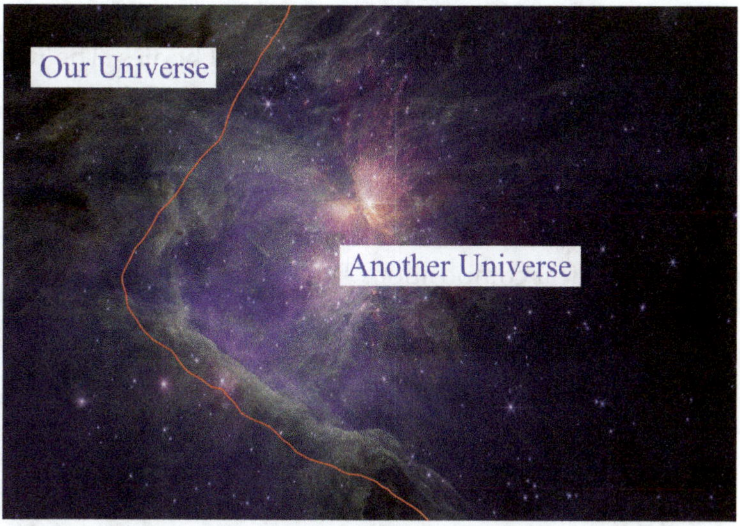

Figure 17.10: The Orion Nebula as seen through the NIRCam instrument. Credit: CSA James Webb Space Telescope, NASA, ESA.)

This visual spectacle is highlighted by ionized purple gas at the center, with green and darker brown tones at the periphery. The nebula fades to near black on one side, interspersed with thousands of fainter stars, the brightest exhibiting characteristic Webb diffraction spikes.

The inner cavity of this young, one-million-year-old star-forming region is predominantly filled with ionized gas, appearing purple. Surrounding areas display a rich mixture of dust and molecular gas in shades of red, brown, and green. The Bright Bay in the upper left is eroded by the massive central stars of the nebula, forming columns of gas and dust.

According to TDU, this inner cavity is not a celestial body but a segment of a white hole: on the left side is our universe and on the right side is another universe, complete with its own sky and celestial bodies. The red line marks the boundary between the two universes, indicating that the gases visible on our side have leaked into our universe from the other.

Another groundbreaking discovery concerns the nebula's chemical composition. The high-resolution spectroscopic capabilities of the NIRCam instrument identified complex organic molecules, offering tantalizing clues to the origin of life's building blocks in such stellar nurseries.

17.3.3 Carina Nebula (NGC 3372)

The Carina Nebula, among the largest and brightest in the sky, is an expanding cloud of gas and dust, serving as a star formation incubator. New stars emerge from the gravitational collapse of dust and gas.

This space region offers astronomers a laboratory to study star life cycles and the interaction of stellar winds and radiation with the interstellar medium [4].

Located about 7,600 light years from Earth, the Carina Nebula is visible in the southern constellation of Carina. It has captivated astronomers for years, with each technological advance revealing more of its mysteries.

The Carina Nebula comprises regions with unique characteristics. A notable feature is the Homunculus Nebula, surrounding the unstable massive star Eta Carinae—one of the most massive binary star systems known. This star often experiences violent outbursts, shedding mass and shaping the nebula's appearance.

Figure 17.11: Carina Cliffs from the Webb Space Telescope. Credit: NASA, ESA, CSA, STScI

The nebula also includes the Keyhole Nebula, a smaller dark nebula of cold molecules and dust, positioned in front of the larger nebula. It absorbs light, creating a stark contrast. Named for its keyhole-like shape from Earth's perspective, it contains dense clumps of material where new stars form.

The varied landscape of the Carina Nebula features vast hydrogen clouds glowing brightly due to ionizing radiation from young, hot stars. These clouds, known as H II regions, consist of ionized hydrogen gas. The mix of dark dust clouds and bright gas creates a striking visual contrast, illustrating the nebula's dynamic processes.

These processes include stellar winds—streams of charged particles ejected from stars' upper atmospheres. Traveling at millions of kilometers per hour, these winds collide with the nebula, creating shock waves that compress gas and dust, leading to new star formation. This cycle of star birth and death is observable throughout the nebula.

The stars of the open star cluster NGC 3324, within the Carina Nebula, significantly influence their surroundings. Their intense radiation and winds shape the nebula's dramatic landscape. Stellar winds remove lighter elements, allowing denser regions to collapse under gravity and form new stars.

The Carina Nebula exemplifies radiation pressure effects. Intense light from massive stars pushes surrounding dust and gas, forming intricate nebula structures. This interplay of forces creates stunning sights like towering pillars and giant voids.

The James Webb Space Telescope (JWST) has focused on this environment, providing an unprecedented view of the nebula. Infrared observations penetrate the dusty veil, revealing hidden star formation processes and intricate structural details. The telescope's advanced instruments capture images with clarity and detail far surpassing previous observations.

The images from JWST reveal the jagged cliffs of the nebula, shaped by the forces of the stars in NGC 3324. These cliffs belong to the Gabriela Mistral Nebula, named after the Chilean poetess due to its silhouette resembling her profile. The telescope's infrared capabilities have unveiled these structures in a new light, exposing previously hidden details.

The Carina Nebula, with its dynamic environment and stunning beauty, remains a captivating object of study in our galaxy. Observations by the James Webb Space Telescope highlight humanity's enduring quest to explore and comprehend the cosmos. As we continue to gaze at the stars, the Carina Nebula reminds us of the wonders awaiting us in the vastness of space.

Comparing the description with the image in Figure 17.11 on page 163, we see that cosmologists termed this celestial phenomenon a "nebula" because they could not conceive it as a white hole into another universe, as the TDU explains. The lower part of the image is nearer to us than the upper part. What they perceive as cliffs is the irregular edge of the white hole.

17.3.4 Ring Nebula (M57)

The fourth celestial body is the Ring Nebula (M57), located in the constellation Lyra and part of the Milky Way galaxy. It is one of the most renowned planetary nebulae, situated about 2,283 light years from Earth.

(a) Original image. (b) Enlarged center.

Figure 17.12: TM57 Ring Nebula.
Credit: NASA, ESA, CSA, JwstKong

According to current knowledge, this nebula is formed by the outer layers of a dying star, leaving a white dwarf at its center. The nebula's colorful rings consist of gas and dust ejected during the star's late evolutionary stages.

Strands of gas resembling eyelashes around a cosmic eye are visible in this digitally enhanced image with assigned colors around the ring.

The nebula introduces elements such as carbon, oxygen, and nitrogen into the interstellar medium. This material will later contribute to the formation of new stars and planets.

However, the Ring Nebula holds a secret that requires knowledge of the multiverse's existence. It is not a nebula or another cosmic object as currently assumed, but a large white hole in the universe's boundary. It also leaks energy and matter from the Milky Way into our universe, providing a glimpse into its interior, which contains stars and a sky presumably similar to Earth's.

* * *

Part VII

ULTIMATE SECRETS OF GEOLOGY

18 PLANETARY SCALE

18.1 INTRODUCTION

On the grand scale of reality, the Planetary Scale lies between the microscopic and cosmic scales. This includes the Solar System, with a focus on Earth and its geological and natural phenomena like earthquakes, volcanic eruptions, landslides, tornadoes, hurricanes, climate patterns, and meteorological events. We demonstrate that the expansion process extends to the Solar System level and provides evidence supporting Earth's expansion.

Earth is a complex and active geosystem with interconnected components interacting dynamically:

Atmosphere: The gas layer surrounding Earth, crucial for regulating climate and weather.

Hydrosphere: Earth's water system, including oceans, lakes, rivers, and groundwater, influencing climate, weather, and geological processes.

Lithosphere: Earth's outer solid layer, consisting of the crust and upper mantle, broken into large plates floating on the fluid mantle below.

Biosphere: The life zone on Earth, encompassing all living organisms, from bacteria to complex ecosystems, interacting with other geosystem components.

18.2 KEY PROCESSES DRIVING EARTH'S GEOSYSTEM

Several key processes drive the interactions within Earth's geosystem, including:

Plate Tectonics: The movement of lithospheric plates, causing earthquakes, volcanic eruptions, and the formation of mountain ranges.

Weather Patterns: Air movement in the atmosphere influences climate, weather, and global heat distribution.

Geological Processes: Processes shaping Earth's surface, including erosion, sedimentation, and metamorphism.

Biological Processes: Interactions between organisms and their environment, affecting ecosystems, climate, and geological processes.

18.3 INTERACTIONS AND FEEDBACK LOOPS

Earth's geosystem components interact through complex feedback loops, amplifying or dampening effects.

For instance, the atmosphere affects climate, influencing global heat distribution and weather patterns.

The hydrosphere regulates climate, with oceans as major carbon sinks, affecting cloud and precipitation formation.

The lithosphere interacts with the atmosphere and hydrosphere through geological processes like volcanic eruptions and mountain formation, impacting climate and weather.

The biosphere interacts with other geosystem components, with organisms affecting ecosystems, climate, and geological processes through photosynthesis, respiration, and decomposition.

18.4 CONCLUSION

Earth is a complex, active geosystem with interconnected components interacting dynamically. Understanding these interactions and feedback loops is vital for resource management, climate change mitigation, and ensuring ecosystem sustainability.

However, the fundamental physical process, "global expansion," which underlies all other natural processes, remains officially unacknowledged.

* * *

19 EARTH HISTORY

The Earth is very old, 4 1/2 billion years or more according to recent estimates. This vast span of time, called geologic time by earth scientists and believed by some to reach back to the birth of the Solar System, is difficult if not impossible to comprehend in the familiar time units of months and years, or even centuries. How then do scientists reckon geologic time, and why do they believe the Earth is so old? A great part of the secret of the Earth's age is locked up in its rocks, and man's centuries-old search for the key led to the beginning and nourished the growth of geologic science [76].

—William L. Newman}

19.1 UNCHANGING EARTH

The notion of an unchanging earth has long been ingrained in human thought, yet recent discoveries challenge this belief. Approximately 200 years ago, James Hutton, a pioneer of modern geology, introduced the "Theory of the Earth" in 1795. Hutton's research showed that ancient rocks formed as sedimentary layers on the ocean floor, transforming into sandstone and shale under pressure. Evidence of buckled and twisted layers indicated movement of the earth's surface. These folds crumpled into steep crustal formations, some rising above sea level. Weathering began on these exposed rocks, eroding their surfaces over vast periods.

Subsequent submersion led to new horizontal layers of sand and mud, forming an angular unconformity. Hutton's findings highlighted the earth's immense age, as weathering required significant time.

This concept of an ancient earth revolutionized geology, allowing humanity to envision the earth's constant change over vast periods. As continents eroded, new sedimentary layers formed. Land levels repeatedly sank and rose, creating a slow cycle of natural events over millions of years. To Hutton, this indicated the earth was immeasurably old, with no beginning or end. Today, radioactive decay dates rocks, with the oldest in Greenland over 3,800 million years old.

These ideas significantly influenced fossil understanding. As the concept of an ancient, ever-changing earth gained acceptance, the existence of animals and plants from those times emerged. Rocks containing sea creature fossils were once beneath ancient seas, but as land rose, fossils became land creatures and plants. This recurring process led to the realization that such transformations occurred repeatedly.

19.2 NEW THEORIES ABOUT EARTH

In the early 1900s, theories about the Earth resembled modern ones but had differences. The Earth was thought to be a solid sphere with constant continents, connected by disappearing land bridges. Renowned scientists defended this theory, but it was incorrect. A few proposed a modernized view.

In 1910, F.B. Taylor showed the northern hemisphere's movement from the poles towards the equator, causing North America to separate from Greenland. He didn't reassemble the continents.

In 1911, Baker proposed a single continent, illustrated with a "displacement globe." Between 1912 and 1914, he published "The Origin of Continental Forms," suggesting a single continent that split into today's continents. The northern continents fit tightly, and the Mediterranean basin is closed by the rotation of Italy, Spain, and Newfoundland, bringing North America to North-west Africa and South America. These land masses were once connected. If reconstructed on a globe the same size as the present Earth, the continents won't fit closely, and Baker's map distorts them significantly.

In 1912, German meteorologist Alfred Wegener proposed the theory of continental drift, independently of previous work. He presented evidence that the continents were once joined in Pangaea 300 million years ago and broke into their current positions.

Wegener's theory was sparked by observing that continent outlines fit like an oversized jigsaw puzzle. Francis Bacon noted this in 1620, and Snider observed geological similarities in 1858, explaining fossil plant similarities in

European and North American coal deposits. However, the idea was considered outrageous and forgotten until Wegener sought further evidence to prove or disprove continental drift.

Wegener's concept of continental drift emerged in 1910 while studying world maps. Initially skeptical, he later found supporting evidence after learning of prehistoric animal and plant similarities. By 1912, he presented his findings in scientific journals and published a book on continental origin in 1915. Revised editions followed, with the fourth edition incorporating evidence from other researchers.

The land bridge theory, prevalent at the time, explained the similarity of ancient fossils across continents. However, Wegener corrected this misconception by highlighting the fundamental differences between land and continents. Comparing continent and ocean floor heights, he revealed two distinct mean elevations: the abyssal plain and the average continental level of 4800 meters above it. Wegener emphasized the clarity and reliability of this discovery in geophysics.

If the oceans and continents were made of the same material, a graph of elevations would peak in a standard Gaussian curve, following probability laws. In practice, there are two peaks: one for continents and one for sea floors. This shows continental blocks are lighter than denser ocean floors. Wegener deduced these blocks floated on the ocean floor. While this concept is still debated, one reason the continental drift theory was rejected was the difficulty imagining continents floating on the ocean floor. The ocean floor is solid rock, so how could continents float? The answer lies in the assumption that rocks are solid over long periods. This is an illusion; rocks tend to flow like liquids over long periods.

Once Wegener demonstrated the fundamental difference between continental blocks and the deep ocean floor, it became implausible for the ocean floor to form continental land bridges. After disproving the land bridge theory, the only explanation for the similarity of ancient fossils was that the continents were once united in Pangaea. The drifting continents theory explains why ancient fossils like earthworms, snails, and plants were nearly identical on distant continents. Since these creatures couldn't have flown or swum across vast oceans, the continents must have been connected.

Fossil evidence shows that life forms unable to cross vast oceans existed on all southern Pangaean continents. For example, Mesosaurus, a small reptile, lived in both Brazil and South Africa during the Paleozoic era. The Glossopteris flora, a tree-like plant with pea-sized seeds, was also present on all southern continents. Additionally, more modern mammal-like reptiles inhabited most southern continents at that time.

Although Wegener provided substantial evidence for continental drift, he never explained why continents moved such vast distances. He was unconcerned by this, stating, "The Newton of drift theory has not yet appeared. His absence need cause no anxiety; the theory is young and still treated with suspicion."

The drift theory lacked a mechanism to explain continental drift, leading to its incomplete acceptance. Despite heated debate until the late 1930s, interest waned after Wegener's death in Greenland. New evidence finally confirmed the theory in the late 1950s.

O.C. Hilgenberg offered an explanation three years later in his 1933 book "All about the growing world." He proposed that continents formed from a continuous shell and drifted apart as the Earth expanded.

While most scientific opinion opposed continental drift, notable supporters like Du Toit advocated for it. Du Toit argued that continental movement was essential for interpreting Paleozoic fold systems in the Southern Hemisphere. He highlighted the revolutionary nature of the hypothesis, even without understanding its causes.

19.3 CONTINENTAL DRIFT REDISCOVERED

After World War II, echo sounding equipment began studying the ocean floor. In the 1950s, data from over 50 expeditions was mapped into "The Floors of the Oceans" (1959) by Bruce Heezen, Marie Tharp, and Maurice Ewing. The map revealed three major ocean floor regions: the continental margin, the ocean basin floor, and the mid-oceanic ridge.

The continental margin includes areas above and submerged under land. These continental seas are shallow and have a steep edge dropping to the ocean floor. The deepest part of the ocean basin floor lies next to the continents, known as abyssal plains, the flattest surfaces of the Earth built up from continental shelves.

Further out, the ocean floor reveals abyssal hills rising from the abyssal floor through scarps until they reach the mid-oceanic ridge, the largest feature on Earth. It's a broad, fractured arch through the ocean floor, rising until it touches the sides of a rift valley, a long cleft 20-35 km wide and 500-1500 fathoms deep. These rift valleys are volcanically active.

Heezen and Ewing compared the mid-Atlantic ridge and African rift valleys, concluding they were similar structures forming part of a continuous feature. The African rift valleys' stretching suggested the ocean floor was splitting apart. Extensive volcanism and intrusion along the ridge could explain this.

In the late 1950s, Heezen proposed how the southern Atlantic formed. Africa and South America broke up by a rift valley similar to the current East African one. Over millions of years, the continents separated, and lava formed a new ocean floor at the center. The ocean floor aged younger towards the center due to this process. Lava continued to well up at the center.

Since the continents joined, the Atlantic Rise aligned with the original rift. Today, it still lies along the continent edges.

19.4 EXPANSION EARTH THEORY

Southern hemisphere geologists, who observed drift first-hand, led the scientific community in accepting it. S. Carey, a Professor of Geology at the University of Tasmania, used physical models and globes to demonstrate the fit between Africa and South America, and he published these findings in works around 1955–1958.

In 1956, Carey organized an international conference on the drift hypothesis and proposed that the earth's expansion caused it. He believed this expansion would help advance the discussion of continental drift. It was in 1981 that he convened a dedicated conference to discuss the drift hypothesis in the context of Earth expansion as the driving mechanism.

However, accepting an expanding earth was challenging. The scale of the problem is evident: if the earth doubled in diameter, its surface area would increase fourfold, and its volume eightfold. One early reason proposed for an expanding earth was the weakening of gravity over hundreds of millions of years. Other factors considered included the earth's heating and possible volume changes due to chemical changes in the mantle.

Pascal Jordan, a professor of physics at the University of Hamburg, proposed a theory to explain the expanding earth. He believed that constant gravity might weaken over time, causing the earth to expand. This theory was first suggested by Dirac in 1937. Jordan discussed the geophysical implications of his theory, including its effects on the earth's long-term climate. He published a German edition in 1966 and an English edition in 1971 to promote his idea.

Jordan argued that Dirac's hypothesis of a weakening gravitational constant explained the earth's expansion. He suggested that higher gravity compresses the earth, but as gravity weakens, the earth expands. This explains the subdivision of the earth's surface into continental shelves and deep-sea basins.

The lowest parts of continents are submerged by the sea, forming continental shelves. Most shelves are narrow, except for the Bering Sea. Jordan believed the ocean floor was created as the Earth expanded, clearly stating his views.

Jordan proposed that gravity's constant weakened over time, causing the Earth's expansion. Hoyle Narlichar (1971, Nature, 233,41) investigated similar ideas, but cosmologists rejected them. A higher gravity constant in the past would have made the sun much hotter, preventing life's evolution. Ice sheet indications suggest a cooler sun.

Scientists' detailed examination of the ocean floor revealed it is progressively younger towards the center. None of the ocean floor is older than about 200 million years, much younger than continental blocks formed over 3800 million years ago."

Research ships recorded the sea floor's rapidly varying magnetic intensity. In 1963, British scientists F. Vine and D. Matthews explained this by proposing that hot lava cooled and recorded the Earth's magnetic field as it spread from a central ridge. Since the Earth's magnetic field reverses periodically, the ocean floor acted as an ancient tape recorder.

Mapping the magnetic recordings and combining them with geological age dating of rock samples revealed that the ocean floor was younger at ridges and progressively older towards continents, with mirror images on each side.

In 1965, Sir Edward Bullard's computer fits of continents demonstrated that the best alignment between South America and Africa used the 2000-meter depth contour. By 1969, Antarctica, Australia, and India were included in this reconstruction. The fit was so precise that continental drift became the only plausible explanation.

All continents had once formed a single landmass that eventually separated.

19.5 EXPANSION THEORY IS REJECTED

Despite overwhelming evidence for continental drift, the Expanding Earth theory was rejected by most scientists due to its astonishing nature. While the Earth was once considered static, the recent discovery of continental movement led to the acceptance of continental drift. However, the expanding earth theory did not gain wide acceptance.

A more conservative explanation proposed a sea floor that renews itself every two hundred million years. The new ocean floor spreads from the central ridge towards the continents, sinks beneath them, and is consumed in the Earth's mantle. This theory, consistent with the traditional view of a constant Earth diameter, was accepted and remains the prevailing theory.

For over 30 years, continental drift was denied due to the belief in solid ocean floors. However, scientific thought shifted to view the ocean floor as part of giant convection cells spanning the globe. Jordan objected, arguing that a heat convection cell needed to overcome internal rock friction. The only energy source was heat at ocean rifts, which must be converted into mechanical work and adhere to thermodynamic laws. From known values, he concluded the system's efficiency was too low to generate convection currents in the Earth's mantle.

When the giant convecting cell theory was proposed in 1960, sea floor knowledge was incomplete, prompting ongoing research to prove or disprove it. These new facts raised unanswered questions.

Other scientists noted the discontinuity of the "ring of fire," a region surrounding the Pacific Ocean where the old crust was believed to be consumed. If the active subduction zones didn't form a continuous line, how could the entire Pacific Ocean floor be subducted? This question remained unanswered but didn't undermine the giant convection cell model.

Despite his inactivity, Carey published articles and organized conferences on the Expanding Earth from 1958 to 1983. In 1983, he edited a collection of contributors. In 1979, Dr. H. G. Owen of the British Museum provided reasons why continental drift could only be reconciled with an Expanding Earth.

Firstly, the precise re-fitting of continents into the Pangaea landmass required a smaller Earth. Secondly, the distribution of land-living Dicynodon reptiles was exceptional if a wide sea separated Greater India from China. These reptiles lived about 200 million years ago in the Triassic period, inhabiting areas of South Africa, South America, Greater India, Antarctica, and China. If a wide sea separated Greater India from China, it was difficult for these animals to migrate between the two land masses. However, with a smaller Earth and the disappearance of the Tethyan Sea, this problem vanished.

The third point was that the North Pacific Ocean supports a constant-diameter Earth concept, but the South Pacific Ocean has vast oceanic crust, requiring an expanding Earth. From the late Cretaceous to the present, new oceanic crust formed around Antarctica, compensating for the lack of subduction areas.

Dr. Owen's book, "Atlas of continental displacement, 200 million years to the present," compares how continents fit on a smaller Earth [81]. It presents two sets of ancient world maps: one depicting a much smaller Earth and the other the current size. This enables a comparison between a constant-diameter Earth and an expanding Earth. With a larger diameter, continents do not fit together, resulting in spherical triangular gaps from the reassembly of Pangaea. As the Earth shrinks, the increasing surface curvature reduces these gaps until they vanish.

Despite the scientific consensus against the Expanding Earth theory, there are reasons to consider it. Unlike giant convection cells, it elegantly explains sea floor spreading and continental movement. The absence of a sea floor older than a few hundred million years and the need for mythical oceans to fill gaps are addressed in the next chapter "Earth Expansion Hypothesis."

* * *

20 EARTH EXPANSION HYPOTHESIS

As a geologist, I insist that the Earth has expanded, and leave it as a cosmological problem of the whole universe [22].
—S. Warren Carey

20.1 EARTH'S EXPANSION AS A NATURAL PHENOMENON

The *Earth Expansion Hypothesis* suggests that Earth's expansion is a natural process marked by a continuous increase in its mass and diameter. As explained in Chapter 15, *Cosmhole Hypothesis,* on page 133, this expansion results from a constant influx of magma from the white mouth in Earth's inner core center.

As magma accumulates, the Earth's crust, formed from previous expansions, must adapt by cracking and releasing pressure at its weakest points, causing earthquakes. The quake's intensity correlates with the pressure required to break the mantle.

Once the pressure of the incoming magma is released, the crust stabilizes until new pressure builds, necessitating a new fissure. Understanding this cycle requires a planetary perspective of the Earth's crust. With technology that measures pressure at various crust points, major future earthquakes could be predicted. However, on average, there are thousands of earthquakes of all magnitudes every day worldwide.

Earth Expansion Tectonics has demonstrated Earth's continuous growth. In his book "Earth Universe Cosmos," S. W. Carey wrote:

> *"The problem is, what triggers the process. As a geologist, I insist that the Earth has expanded, and leave it as a cosmological problem of the whole universe [22]."*

S. W. Carey is correct: Expansion is fundamentally a cosmological issue. Like all celestial bodies, Earth has a white hole at its core's center. See Appendix for more on S. W. Carey.

20.2 EARTH EXPANSION PHASES

- Approximately 12.3 billion years ago, the space where Earth now exists was empty.
- Around 12.2 billion years ago, the first drop of magma emerged from a new white hole, only 0.001 mm in diameter. As magma continued to flow, the sphere expanded, reaching a diameter of 1 meter in a billion years.
- When Earth was 3.7 billion years old, its diameter was 1 km. Initially, Earth's temperature matched the incoming magma. As its diameter increased, Earth's surface radiated more energy into space, cooling and becoming less fluid.

- About 4,500 million years ago, with a diameter of 50 km, Earth's surface solidified. The crust withstood internal pressure from the magma for about 2 billion years. This moment often determines a celestial body's future.

- If the crust withstands the pressure, no new magma enters, and the sphere gradually cools to 2.725 K, the universe's temperature as measured by the cosmic microwave background radiation (CMB).

- Fortunately, Earth's crust could not withstand the internal pressure, leading to surface cracks and continued growth. Each crust break caused slight tremors.

- Over billions of years, Earth's solid mantle thickened and cracked. Occasionally, magma coming from the source universe reaches the surface, causing volcanoes. Earth's continuous expansion has led to cracks, earthquakes, and volcanic activity.

- Currently, Earth experiences about 8,000 earthquakes daily, including all magnitudes, even the smallest (down to magnitude 1.0). This clearly indicates that our globe is expanding daily.

The expansion process is detailed in Chapter 12, *Theory of Decaying Universes*, on page 121.

* * *

21 | EARTH EXPANSION HISTORY

Earth expansionists, since the pioneering work of Christopher Otto Hilgenberg in 1933, have known that if all of the Earth's continents were fitted together they would neatly envelope the Earth with continental crust on a globe some 55 % to 60 % of its present size. This coincidence led Hilgenberg to conclude that terrestrial expansion brought about the splitting and gradual dispersal of continents as they moved radially outwards during geological time. This coincidence has, however, consistently failed to gain recognition within the scientific community as a viable explanation for modern global tectonics. The primary reason for this, in addition to palaeomagnetic-based conclusions, is considered to have been the general lack of supportive global scale evidence to quantify accurately a reproducible expansion process with time [68].

—James Maxlow

Since the 1800s, astute geologists gradually started to discover that the Earth is expanding, but could not provide at the same time any reasonable explanation or hypothesize on a verifiable mechanism of expanding.

Charles Lyell, a Scottish geologist, who was Charles Darwin's mentor, is known as the author of *Principles of Geology* and for explaining the earth's history through known natural causes. He was the first to suggest the existence of forces acting near the Earth's crust on smaller areas.

Darwin made a step further when he found out that a huge area of South America had been *uplifted to its present height by a succession of elevations which acted over the whole of this space with nearly an equal force [109].*

Darwin also hypothesized that an uplift at such a continental scale required *the gradual expansion of some central mass* [of the earth] *acting by intervals on the outer crust with the elevations being concentric with form of globe (or certainly nearly so).*

Other scientists found also confirmations that the Earth is expanding.

In 1889 and 1909 Roberto Mantovani, a musician and geologist, assumed that one continent covered the entire surface of a smaller Earth, and published a hypothesis of Earth expansion and continental drift [67].

In 1912, Alfred Wegener proposed the theory of *Continental Drift,* which suggested that the continents were joined together at a certain time in the past and formed a single landmass known as Pangaea.

Since the pioneering work of Christopher Otto Hilgenberg in 1933, earth expansionists have known that if all of the Earth's continents were fitted together they would neatly envelope the Earth with continental crust on a globe some 55% to 60% of its present size.

This coincidence led Hilgenberg to conclude that terrestrial expansion brought about the splitting and gradual dispersal of continents as they moved radially outwards during geological time.

Some of these scientists continued and developed Hilgenberg's work. Others dismissed it because of "the lack of a mechanism to explain the expansion." Unfortunately, even Earth expansion geoscientists themselves consider that *...there is little immediate or pressing need (as there was with evolution in biology) to quickly adopt the Earth expansion model. What is beyond doubt ... is that the expansion geotectonic model will be adopted in the next two decades.* As David Ford states, *...Ubiquitous ignorance of Earth expansion will eventually fade as the overwhelming evidence of global expansion becomes more widely disseminated and generally understood.*

* * *

22 EARTH EXPANSION EVIDENCE

22.1 FACTS ABOUT SOLAR SYSTEM AND EARTH

The following facts are among many that are either incorrectly explained or remain unexplained by established knowledge. They all support the expansion hypothesis.

Fact 1

In 1940, Inge Lehmann, a seismologist at the Royal Danish Geodetic Institute, hypothesized that the inner core is composed of solid nickel-iron. She based this on seismic wave data and their travel times through the Earth. This inner core is surrounded by an outer layer that behaves like a fluid [100].

In 1952, Francis Birch published a detailed analysis of the available data on Earth's core and concluded that the inner core was probably crystalline iron.

In 1984, Hugh Owen wrote an article in which he said [82]:

> *"The geological and geophysical implications of such Earth expansion are so profound that most geologists and geophysicists shy away from them. In order to fit with the reconstruction that seems to be required, the volume of the Earth was only 51 % of its present value,*

> *and the surface area 64 % of that of the present day, 200 million years ago. Established theory says that the Earth's interior is stable, a solid. Perhaps we are completely wrong and the inner core is in some state nobody has yet imagined, a state that is undergoing a transition from a high-density state to a lower density state, and pushing out the crust, the skin of the Earth, as it expands."*

The geological and geophysical implications of such Earth expansion are so profound that most geologists and geophysicists shy away from them. In order to fit with the reconstruction that seems to be required, the volume of the Earth was only 51 per cent of its present value, and the surface area 64 per cent of that of the present day, 200 million years ago. Established theory says that the Earth's interior is stable, an inner core of nickel iron surrounded by an outer layer that behaves like a fluid. Perhaps we are completely wrong

Figure 22.1: The earth is expanding and we don't know why

After 37 years, in October 2021, Owen's prediction-that *"perhaps we are completely wrong and the inner core is in some state nobody has yet imagined"* proved correct. The *Nature World News* magazine published the article *Scientists Discovered Unexpected 'Hidden World' in Earth's Inner Core! [105]:*

> *"The more that we look at it, the more we realize it's not one boring blob of iron,* Jessica Irving, a seismologist at the University of Bristol in England, who was not involved in the study, told Live Science: *We're finding a whole new hidden world."*

The researchers discovered that the detected waves suggested the inner core contained pockets of liquid and mushy, semisolid iron at its surface, rather than being entirely solid.

Butler, a co-author, noted:

> *"We've seen evidence that not only is it not soft everywhere; it's really hard in some places. It's got hard surfaces right up against melted or mushy iron. So we're seeing a lot of detail within the inner core that we didn't see before."*

Fact 2

In a new article published on February 10, 2025, scientists discovered that *the inner core's structure changes as it rotates. It deforms at its border, potentially accumulating more material in some areas and less in others—almost like creating hills and valleys* [84].

They also found that *Earth is divided into several layers: crust, mantle, outer core, inner core, and even a recently suspected innermost inner core*, and that *Earth's inner core—like a planet inside another planet—appears to move to its own beat.*

The team of scientists, who previously discovered that the inner core spins at slightly different speeds than the rest of solid Earth, now found that the inner core's structure also changes as it rotates. *It deforms at its surface, the new study says, potentially accumulating more material in some areas and less in others—almost like creating hills and valleys.*

Fact 3

In recent years, accumulating evidence has provided insights into why Earth is expanding.

In 2013, a discovery revealed that *Earth's inner core spins faster than Earth itself* [44].

The authors highlighted the significance of this finding [1]:

> *"Seismic probing of the earth's deep interior has shown that the inner core, the solid core of our planet, rotates slightly faster (i.e., eastward) than the rest of the earth. Quite independently, observations of the geomagnetic field provide evidence of westward-drifting features at the edge of the liquid outer core."*

Scientists investigating this *anomaly* attempted to explain it using a *3D numerical model of the Earth's core with a viscosity two orders of magnitude lower than the state of the art,* a value one hundredth of the actual, which is negligible. However, they could not determine why the inner core influences Earth's rotation.

However, a new 2024 study confirms that Earth's inner core has been rotating more slowly than usual since 2010 [10].

The key point is that the inner core's rotation, whether faster or slower, differs fundamentally from the outer core and the entire Earth.

> *"The heart of our planet has been spinning unusually slowly for the past 14 years, new research confirms. And if this mysterious trend continues, it could potentially lengthen Earth's days—though the effects would likely be imperceptible to us."*

Fact 4

Earth's core is significantly hotter than previously believed.

> *"New measurements suggest the Earth's inner core is far hotter than prior experiments suggested, putting it at 6,000°C, give or take 500°C—as hot as the Sun's surface. The solid iron core is actually crystalline, surrounded by liquid. ... But the temperature at which that crystal can form had been a subject of long-running debate [83]."*

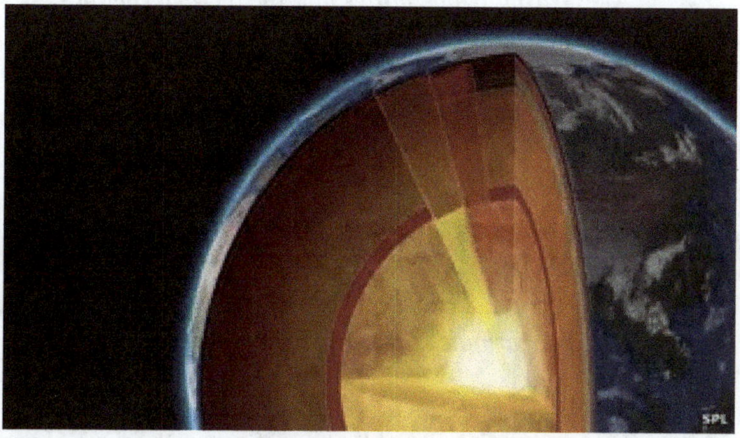

Figure 22.2: Earth's core far hotter than thought. Credit BBC News [83]

Fact 5

Superionic iron alloys and their seismic velocities in Earth's inner core

For over fifty years, scientists thought Earth's center was a molten outer core surrounding a densely compressed solid iron alloy. However, new research in *Nature* reveals a much stranger inner structure than previously assumed.

According to its author [57]:

> *"Earth's inner core (IC) is less dense than pure iron, indicating the existence of light elements within it. Silicon, sulfur, carbon, oxygen and hydrogen have been suggested to be the candidates, and the properties of iron-light element alloys have been studied to constrain the IC composition. Light elements have a substantial influence on the seismic velocities, the melting temperatures and the thermal conductivities of iron alloys. However, the state of the light elements in the IC is rarely considered. Here, using ab initio molecular dynamics simulations, we find that hydrogen, oxygen and carbon in hexagonal close-packed iron transform to a superionic state under the IC conditions, showing high diffusion coefficients like a liquid. This suggests that the IC can be in a superionic state rather than a normal solid state."*

"It is quite abnormal," study first author Yu He, a geophysicist at the Chinese Academy of Sciences, said in a statement. "The solidification of iron at the inner core boundary does not change the mobility of these light elements, and the convection of light elements is continuous in the inner core."

Fact 6

All planets orbit in a flat plane around their suns.

The planets in our solar system orbit our Sun more or less in a single plane within 3 degrees of the ecliptic. Mercury is a bit different because its orbit is inclined to the ecliptic by 7 degrees. Pluto, a dwarf planet, has its orbit inclined to the ecliptic by more than 17 degrees and is widely known to be an exception to this rule [36].

According to the EarthSky editors:

> *"It makes sense that most large planets in our solar system stay near the ecliptic plane. Our solar system is believed to be about 4 1/2 billion years old. It's thought to have arisen from an amorphous cloud of gas and dust in space. The original cloud was spinning, and this spin caused it to flatten out into a disk shape. The Sun and planets are believed to have formed out of this disk, which is why, today, the planets still orbit in a single plane around our Sun."*

Astronomers note that exoplanets orbit distant stars in a similar single plane. However, they do not address a key question: *What drives these suns and disks to spin so rapidly?*

Refer to the image of the young star HL Tau on page 200 and its protoplanetary disk, provided by The National Radio Astronomy Observatory.

According to these scientists, this is the best image of planet formation, revealing multiple rings and gaps that indicate emerging planets clearing their orbits of dust and gas [20].

Another question arises: What if these disks are not real, but technical artifacts of planets spinning too fast around their suns relative to the long exposure time needed for the images? This could explain the gap in the image, not as *heralding the presence of an emerging planet*, but as a lack of a planet in that location.

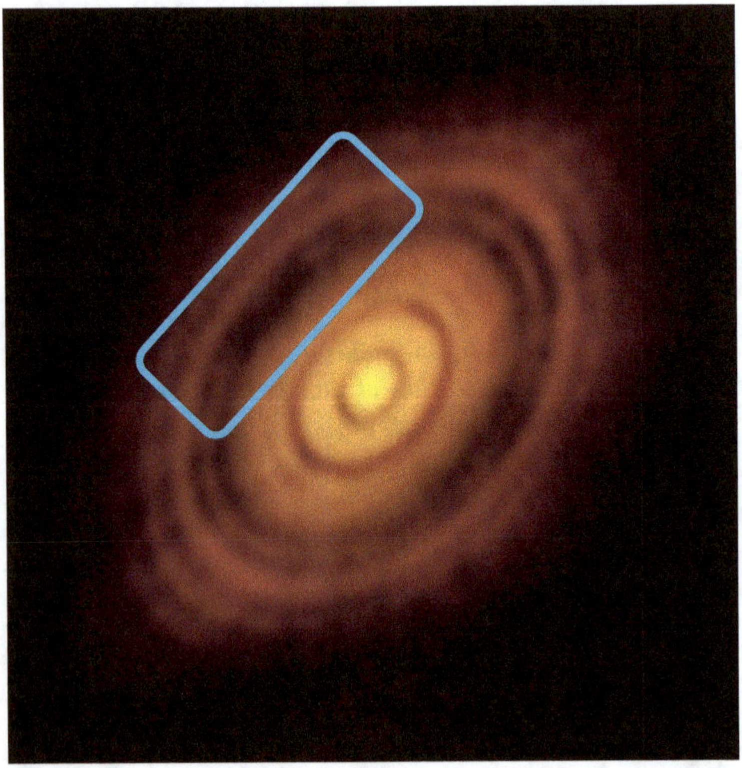

Figure 22.3: HL Tau: Birth of Planets. Credit: ALMA (ESO/NAOJ/NRAO); C. Brogan, B. Saxton (NRAO/AUI/NSF); CSC CC BY-NC-ND 4.0

Fact 7

Solar Rotation Varies by Latitude

The Sun is known by everyone to be *the star at the center of the Solar System. It is a nearly perfect ball of hot plasma, heated to incandescence by nuclear fusion reactions in its core. The Sun radiates this energy mainly as light, ultraviolet, and infrared radiation, and is the most important source of energy for life on Earth [8].*

Also, *roughly three-quarters of the Sun's mass consists of hydrogen (73 %); the rest is mostly helium (25 %), with much smaller quantities of heavier elements, including oxygen, carbon, neon, and iron [11].*

As highlighted by NASA, the Sun is not a solid ball, as we would suspect. On the contrary:

> *"The Sun rotates on its axis once in about 27 days. This rotation was first detected by observing the motion of sunspots... Since the Sun is a ball of gas/plasma, it does not have to rotate rigidly like the solid planets and moons do. In fact, the Sun's equatorial regions rotate faster (taking only about 24 days) than the polar regions (which rotate once in more than 30 days). The source of this "differential rotation" is an area of current research in solar astronomy [125]."*

Scientists have concluded that the Sun's mass consists of hydrogen and helium, heated by nuclear fusion. If this is accurate, what causes the Sun to spin? At least one data set—either the Sun's composition or the heating source—must be Incorrect. Or worse, perhaps both are wrong.

The Earth's inner core temperature is comparable to the Sun's, both approximately $6,000°C \pm 500°C$.

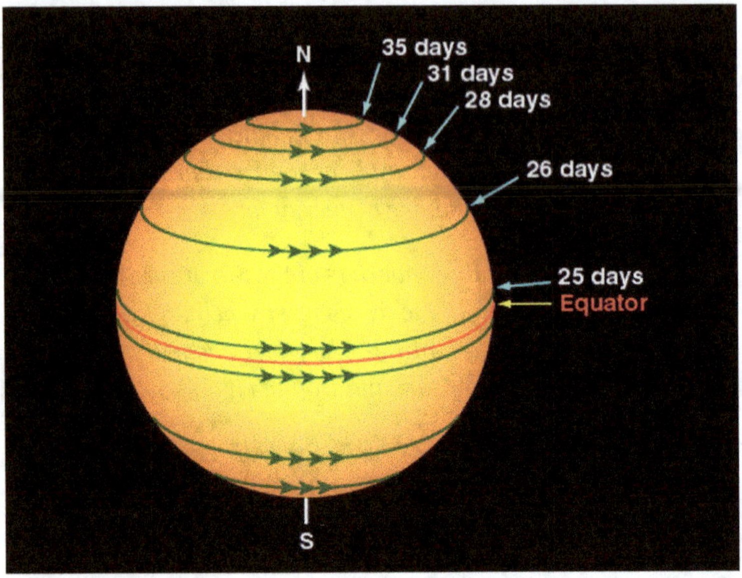

Figure 22.4: Solar rotation varies by latitude.
CREDIT: Holly Zell, NASA

This similarity suggests a shared type of heating source in their cores. The distinction is in the fluidity of the material's incoming flow.

Corroborating with the fact that the Sun's polar regions lag behind the equatorial regions suggests that the actual heating source is within the Sun's core. The material expands more toward the equatorial regions due to centrifugal force than toward the polar regions.

The Sun's core receives a high flow, maintaining its extreme heat, while the Earth receives a lower flow, heating only its inner core to the Sun's temperature. Despite being small, the accumulation of new matter causes the Earth to expand slowly and steadily.

Fact 8

> It is found that *the spin of low-mass galaxies is more likely to be aligned with the filaments of the cosmic web and to lie within the plane of the walls while more massive galaxies tend to have a spin perpendicular to the axis of the filaments and to the walls [2].*

Carey's insight was recognizing Earth's genesis as a cosmological phenomenon. To complete this understanding, we must also consider the insights of Sir Martin Rees, an astrophysicist, cosmologist, and the UK's Astronomer Royal.

22.2 EXPANDING ERODING EARTH

In paragraphs *Fact 1* and *Fact 2* of section 22.1 on page 193, we determined that the suspected innermost inner core resembles a planet with a heteroclitic composition. Consequently, Earth's entire structure, including its crust, shares this diverse composition, consistency, and surface thickness. Therefore, we should expect cracking under expansion pressure to manifest in various ways.

By viewing Earth through the "expansion lenses," we can explain the structure of any location and size on the globe, from continents to small $10m^2$ areas, through two ubiquitous phenomena: expansion and erosion.

These phenomena are more common than one might think; they surround us even in familiar landscapes. Yet, we rarely ask ourselves: "Why are they like that?"

The optimal method to illustrate Earth's relief is through examples of expansion and erosion in action.

22.2.1 Typical Expansion

Figure 22.5: Stages of a river's growth.
Credit "Atlas of the World" (1990)

The above image is reproduced on page 204 from the "Atlas of the World" and depicts the stages of a river's growth [19].

In the picture, the river becomes visible at #2, forms a waterfall at #3, and flows through #4, #5, until down to #11. The riverbed's width remains relatively constant, but between #3 and #5, it flows through a canyon much wider than the riverbed.

Official geology attributes the canyon's formation to water erosion. However, there's a significant discrepancy between the river's size and the canyon's size.

The expansion hypothesis suggests the canyon resulted from local Earth expansion in this area, representing typical Earth expansion.

Initially, the land was continuous. As the Earth expanded, it widened a small crack in the crust, allowing groundwater to flow. In the mountainous region, the crust cracked slightly, causing minimal separation. The river formed as the banks gradually moved apart, with the thinning crust promoting land expansion.

22.2.2 Smallest Expansion After Izmit Quake

In 1999, Tom Fumal, a geophysicist with the US Geological Survey's Western Earthquake Hazards Team, captured the image in Figure 22.6 on page 206 to illustrate Earth's deformation following the 1999 Izmit quake.

To highlight Earth's expansion, in Figure 22.7 below, we marked the photo with lines indicating the expanded area in both directions. The two red arrows show the fault's sides, which were connected before the earthquake.

The diagram in Figure 22.8 on page 208 shows the tree line before and after the earthquake. Initially, the tree line was continuous. After the earthquake, the small square in the left image expanded, shifting the tree line 3.3m to the left and separating the trees by about 3m. (Refer to the position of the two red arrows before and after the earthquake.)

Figure 22.6: Local earth expansion
Credit: U. S. Geological Survey

22.2.3 Earth Expansion Around Greater New York

Refer to Figure 22.9 for a map showing expansion areas of different sizes and shapes.

22.2.4 Earth Expansion Around Greater Montreal area, Canada

This area showcases successive expansions, including a significant one that separated the Montreal and Laval isles from the two shores, and several smaller ones, such as the Lake of Two Mountains.

Refer to Figure 22.10 to view this map with expansion areas of various sizes and forms.

Figure 22.7: Lines highlighting the expanded area. Credit CSC CC BY-NC-ND 4.0.

22.2.5 Earth Expansion Around Berlin, Germany

View this map in Figure 22.11 on page 211 with several expansion areas where the plate is thicker.

22.2.6 Earth Expansion Around Melbourne, Australia

Refer to Figure 22.12 on page 212 to view the map showing expansion areas of different sizes and shapes.

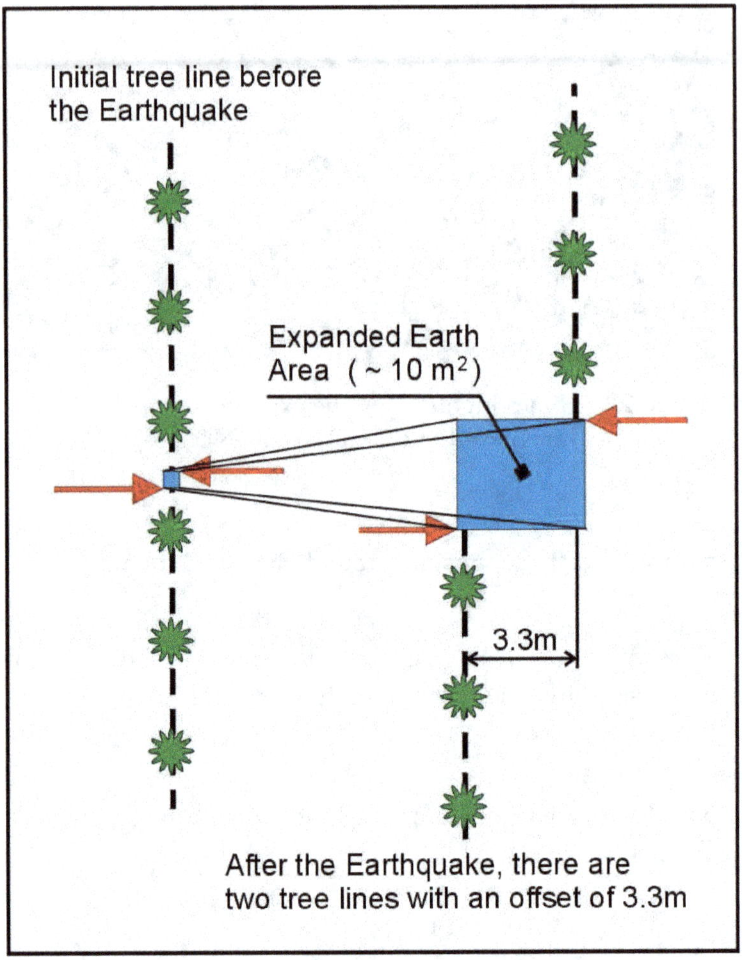

Figure 22.8: Diagram showing the local expansion
Credit CSC CC BY-NC-ND 4.0.

Figure 22.9: Earth expansion around Greater New York, USA. Credit CSC CC BY-NC-ND 4.0.

22.2.7 Earth Expansion Examples in Japan

View this map with expansion areas of various sizes and forms in Figure 22.13 on page 213.

* * *

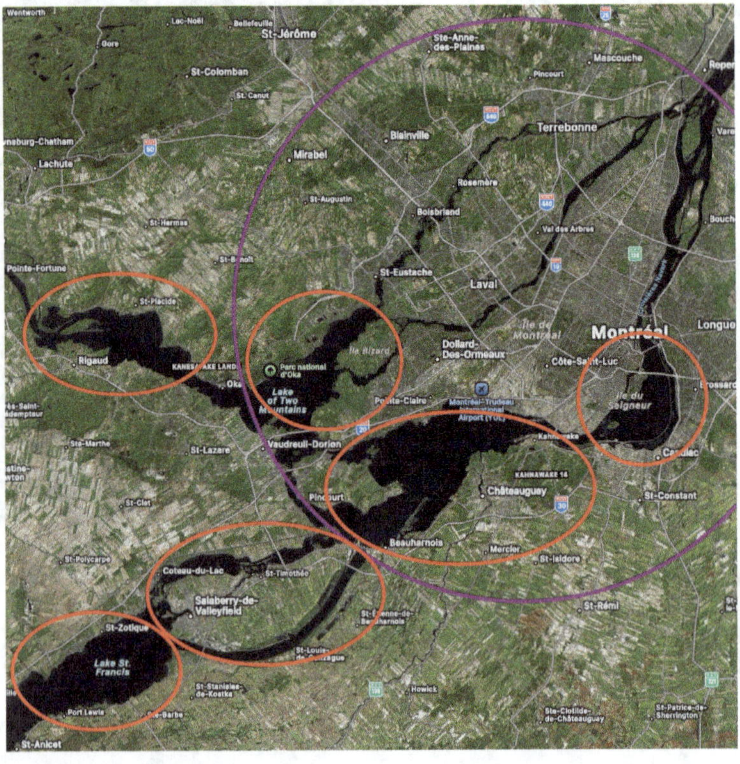

Figure 22.10: Earth expansion around Greater Montreal, Canada. Credit CSC CC BY-NC-ND 4.0.

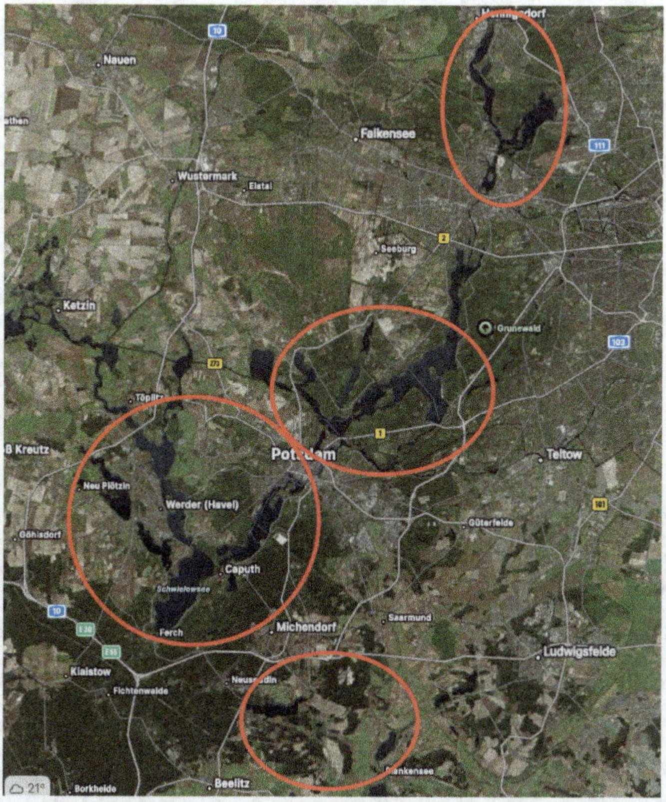

Figure 22.11: Earth expansion around Berlin, Germany Credit CSC CC BY-NC-ND 4.0.

Figure 22.12: Earth expansion around Melbourne, Australia
Credit CSC CC BY-NC-ND 4.0.

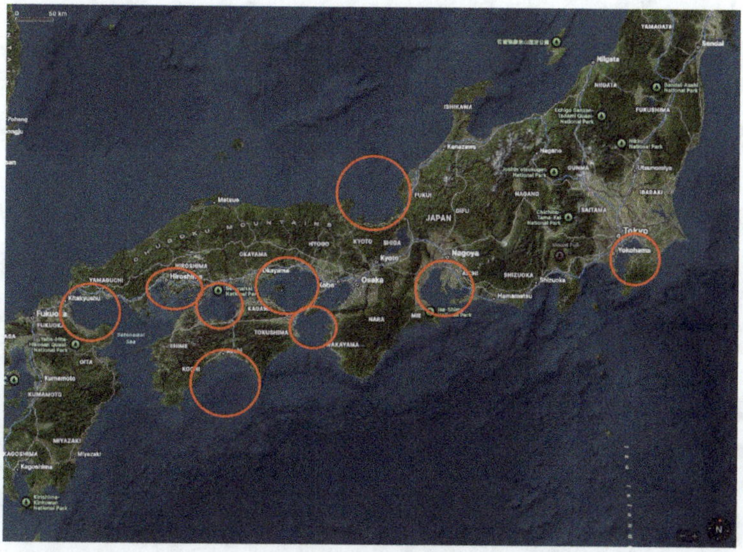

Figure 22.13: Earth expansion examples in Japan
Credit CSC CC BY-NC-ND 4.0.

23 EARTHQUAKES: EVIDENCE OF EXPANSION

If . . . (one)'s explanation of the expansion is found to be inadequate, this would only mean that another explanation must be developed for as with evolution, there remains no other scientific explanation possible for what is observed.

—David Ford

Not all Earth expansions cause earthquakes, but many do, depending on various factors, including the depth and type of rocks involved in cracking.

Earth expansions can be categorized into three main types based on the extent and form of the cracks they create on the Earth's surface:

23.1 SMALL SURFACE CRACKS

These cracks are observed on roads or other relatively small surfaces.

Here are a few examples:

Figure 23.1: Small surface cracks #1.

Figure 23.2: Small surface cracks #2.

Figure 23.3: Small surface cracks #3.

Figure 23.4: Small surface cracks #4.

23.2 MEDIUM SURFACE CRACKS

These cracks are large enough to cause buildings to collapse into them.

Here are some examples:

Figure 23.5: Medium surface cracks #1.

Figure 23.6: Medium surface cracks #2.

Figure 23.7: Medium surface cracks #3.

Figure 23.8: Medium surface cracks #4.

23.3 LARGE CRACKS AFTER MULTIPLE EXPANSIONS

These are the largest global and multiple expansions, most visible in canyons. On the Earth, there are approximately 9,000 canyons.

Here are examples of expansions followed by erosion:

Figure 23.9: Large cracks after multiple expansions #1.

Figure 23.10: Large cracks after multiple expansions #2.

Figure 23.11: Large cracks after multiple expansions #3.

Figure 23.12: Large cracks after multiple expansions #4.

Part VIII

PROGNOSIS FOR EARTH'S FUTURE

24 EARTH'S FUTURE

In one billion years, Earth will transform into a Sun, and the Sun will evolve into a red supergiant star, like Betelgeuse in the Orion constellation.

—Author

24.1 FOUNDATION OF OUR PREDICTION

Our tentative prediction of Earth's future evolution relies on the following assumptions made by other scientists and us:

- S.W. Carey suggested that Earth has expanded, proposing a proto-Earth with a radius of about 100 km;
- Earth Expansion is a continuous process without abrupt changes;
- Recent USGS estimates indicate Earth is over 4.5 billion years old [76];
- 4.5 billion years ago, the proto-Earth had a diameter of about 55 km, a size that still allows geologists to find and date rock samples forming Earth's crust;
- Several authors claim Earth's diameter doubled in the last 200 million years;

The equation that satisfies this growth assumption is:

$$D = D_0\, e^{(GT)^3} \qquad (24.1)$$

Where:

- D is the current Earth's diameter in Km;
- D_0 is Earth's diameter at the reference point, 4.5 billion years ago: 55 Km;
- G is the Average Growth Factor (AGF), calculated as 3.90761E-04 for Earth;
- T is Time, in millions of years, since Earth's diameter D_0 was 55 Km, 4,500 million years ago;
- e is the base of natural logarithms, approximately 2.71828.

The Average Growth Factor G depends on several factors:

1. On the source universe side:
   a) On the size and rotational speed of the vortex originating the cosmhole;
   b) On the composition and fluidity of the heterogeneous magma encountered by the traveling vortex;
2. On the receiving universe side:
   a) The type of white hole that emerged from the cosmhole: C-white hole or O-white hole;
   b) In the case of a C-white hole, on the cooling speed: LPC-white hole or HPC-white hole.

We named equation 24.1 the *Earth Expansion Equation,* derived under the assumption that the expansion process occurred in five phases. (Refer to the chapter on the Cosmhole Hypothesis on page 133.)

24.2 EARTH'S GROWTH FORECAST

Prediction:

Within the next billion years, Earth will transform into a Sun, and the Sun will evolve into a red supergiant star.

The graph on page 229 shows Earth's evolution from 4.5 billion years ago until it exceeds the Sun's size in one billion years.

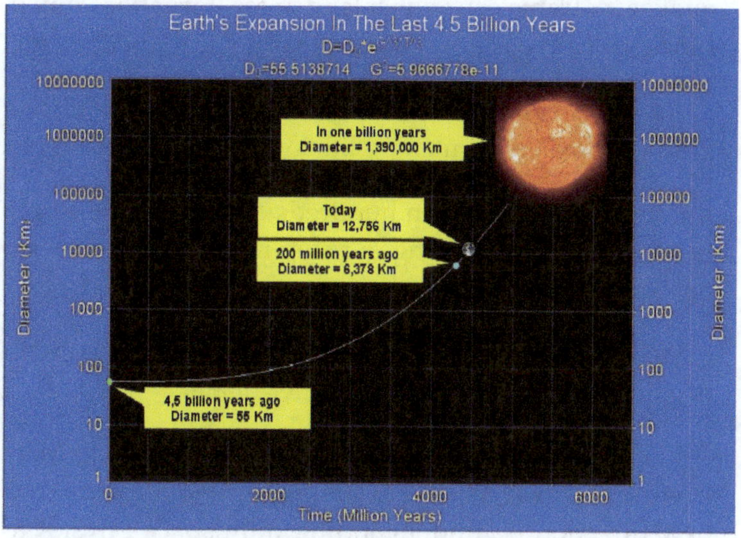

Figure 24.1: Earth's size growth over six billion years.

We conclude that the Earth's actual age is approximately 11 billion years. This differs from the 4.54 billion years estimated for the Solar System and Earth but aligns with current calculations of 11 to 13 billion years for the Milky Way's age.

This estimation fully complies with the USGS-supported hypothesis that the Universe's age is between 10 and 15 billion years.

24.3 IMPORTANT CAVEAT

It is important to remember that Earth is an active geosystem; it is not merely expanding. This process has multiple side effects we know but do not fully understand: earthquakes, volcanic eruptions, landslides, tornadoes, hurricanes, climate patterns, and meteorological phenomena.

As Earth expands, the intensity of these side effects gradually increases, making life increasingly challenging.

Eventually, perhaps in hundreds of years, we will face Earth's deadliest challenges: rising mega earthquakes on one side and extreme global warming on the other.

We cannot predict which of the two will leave us with no choice but to relocate to another planet.

Elon Musk's intuition about moving humanity to Mars is correct. However, his technology must advance in a new direction by understanding multispace physics to develop the appropriate means.

* * *

Part IX

ULTIMATE SECRET OF HUMANITY

25 CENTRAL CHALLENGE TO PHYSICALISM

"There can be no doubt that consciousness does not occur in physics and chemistry, and I cannot see how it could possibly result from quantum mechanics [58]."

—Werner Heisenberg

It may at first appear that the central challenge for physicalism—the mind–body problem—is not physical in nature and therefore falls outside the concerns of this inquiry. Yet it is precisely by unveiling this "most" ultimate secret of reality that all other physical ultimate secrets become comprehensible.

25.1 ON PHYSICALISM

Physicalism, the present paradigm of reality, is the thesis that everything is physical, or as contemporary philosophers sometimes put it, that everything supervenes on, or is necessitated by, the physical. The thesis is usually intended as a metaphysical thesis, parallel to the thesis attributed to the ancient Greek philosopher Thales, that everything is water, or the idealism of the 18th Century philosopher Berkeley, that everything is mental.

The general idea is that the nature of the actual world (i.e. the universe and everything in it) conforms to a certain condition, the condition of being physical. Of course, physicalists don't deny that the world might contain many items that at first glance don't seem physical—items of a biological, or psychological, or moral, or social nature. But they insist nevertheless that at the end of the day such items are either physical or supervene on the physical [108].

This perspective asserts that everything in the universe—matter, energy, forces, and mental phenomena—can be fully explained by physics. Rooted in ancient materialist philosophy and bolstered by modern scientific advancements, physicalism rejects the existence of independent non-physical substances, such as Cartesian souls or vital essences. It advocates a monistic framework where higher-level properties emerge from or depend on physical foundations.

While acknowledging that *many items that initially don't appear physical—such as biological or psychological items, etc.* and simultaneously insisting that *such items are either physical or supervene on the physical,* physicalism presents numerous challenges across various sciences. In previous chapters, we addressed these challenges in Mathematics, Physics, Cosmology, and Geology.

Now, it is time to address the most astonishing challenge for every human being on this planet.

25.2 MIND–BODY PROBLEM.

The principal challenge confronting physicalism is the hard problem of consciousness, long situated within philosophy as the core of the mind–body problem. This problem underscores the ontological and explanatory difficulty of accounting for how subjective mental phenomena—such as consciousness, intentional thought, and affective experience—arise from the physical structures and processes of the brain and body.

Physicalist approaches have tried to address this issue using reductionist identities [90, 104], supervenience relations [30, 61], or emergentist frameworks [18, 26]. However, none has fully explained how first-person experience arises from third-person physical descriptions.

Historical precedents for this dichotomy trace back to ancient thinkers. Plato, in works like Phaedo, viewed the soul (mind) as an eternal, immaterial Form imprisoned in the ephemeral body, striving for liberation through reason. Aristotle refined this with a hylomorphic view, where the soul is the form actualizing the body's potential, though he emphasized their inseparability in living organisms, differing from Descartes' radical independence.

In Eastern philosophy, such as the Buddhist Five-Aggregate Model, the mind is depicted as a flux of impermanent sense impressions and mental phenomena, rejecting a fixed dualistic split in favor of interdependent processes.

The dichotomy poses significant challenges, particularly the interaction problem: if mind and body are disparate substances, how do they causally influence one another?

The classic mind-body dichotomy, given its clearest formulation by René Descartes in the 17th century, posits a fundamental distinction between two substances: the immaterial, non-spatial mind (res cogitans), capable of thought, doubt, and self-awareness, and the physical, mechanistic body (res extensa), governed by physical laws. This substance dualism raises the enduring interaction problem—namely, how two such disparate entities could causally influence one another [33].

Contemporary debates often critique this dichotomy as rooted in intuitive psychological biases, such as an innate tendency toward dualism that separates ethereal mental states from tangible bodily ones [13, 14].

The hard problem of consciousness, introduced by philosopher David Chalmers in 1995, specifically addresses the difficulty of explaining why physical brain states are accompanied by subjective, qualitative experiences (known as qualia), rather than merely functional processes without phenomenal awareness [25, 64].

In contrast to "easy problems" of consciousness, which involve mechanistic accounts of cognitive functions like attention or perception, the hard problem questions the emergence of first-person experience from third-person physical descriptions, challenging materialist or physicalist frameworks that reduce all phenomena to brain activity [25]. Some scholars argue that this problem stems from human psychological predispositions toward dualistic thinking, rather than an inherent metaphysical gap, while others maintain it presupposes or reinforces dualism by highlighting irreducible mental properties [13][27].

These concepts are interconnected, as the hard problem amplifies the mind-body dichotomy by underscoring the apparent incommensurability between objective physical explanations and subjective mental realities [64]. For instance, critics of Cartesian dualism suggest that framing consciousness as inherently "hard" perpetuates an outdated separation of mind and matter, potentially overlooking integrated approaches from neuroscience or phenomenology [32].

Resolving this interplay remains a central pursuit in philosophy, cognitive science, and related fields, with ongoing discussions emphasizing the need for interdisciplinary perspectives to bridge the explanatory divide.

Efforts to dismiss the problem as illusory or to propose solutions like panpsychism or illusionism often stray from strict physicalist commitments, suggesting that resolving this issue may require a paradigm shift in our understanding of reality.

25.3 ARE MINDS AND SOULS ALSO PHYSICAL?

In physics, elementary particles, such as electrons and quarks, are modeled in quantum field theory as excitations of underlying fields, often treated as point-like with no intrinsic spatial extent in their fundamental description [121]. In this sense, mainstream physics explains them by being dimensionless, existing within the mathematical structure of quantum spaces, where their properties emerge from probabilistic wave functions rather than classical volumes. However, this dimensionless quality does not imply non-physicality; these particles interact via fundamental forces and are integral to the observable universe.

Simply put, elementary particles are physical, even if they are dimensionless in our physical space. Scientists also hypothesized that minds and souls, which are also dimensionless, can be physical—just like elementary particles—in their own quantum spaces. This inquiry posits that minds and souls, characterized as dimensionless entities, might possess a physical nature analogous to elementary particles within quantum frameworks. This proposition intersects philosophy, physics, and cognitive science, warranting a careful examination of established concepts and speculative theories.

However, by contrast, souls and minds are traditionally regarded in philosophical traditions as non-material or immaterial essences, lacking spatial dimensions. The concept of the mind often aligns with consciousness or subjective experience, while the soul is frequently invoked in metaphysical or religious contexts as an enduring, noncorporeal aspect of identity. The suggestion that they could be physical in a quantum sense draws from theories attempting to bridge quantum mechanics with consciousness.

Certain hypotheses propose that consciousness or the mind arises from quantum processes, such as coherent quantum states in biological systems, potentially rendering it physical albeit non-classical [53]. For instance, some models suggest the soul as a quantum field or collection of quantum-like fields that interact selectively with physical matter, such as electromagnetic waves, rather than classical particles [12, 101, 107].

These fields could exist in superposition states, analogous to quantum particles, implying a dimensionless, wave-like presence in quantum spaces [15]. Proponents argue this framework accommodates phenomena like nonlocality or entanglement, potentially explaining intuitive or spiritual experiences [91, 117].

Nevertheless, such interpretations remain highly speculative and lack empirical consensus. Mainstream physics and neuroscience attribute consciousness to classical neural processes, with quantum effects deemed negligible due to rapid decoherence in warm, biological environments [111]. Critics contend that invoking quantum mechanics for minds or souls introduces unnecessary complexity without testable predictions, potentially conflating metaphysical concepts with physical laws [62].

Before this book, no experimental evidence supported the existence of souls as (quantum) entities, and dimensionless qualities alone were insufficient to equate them with elementary particles.

* * *

26 | ULTIMATE SECRET OF HUMANITY

"The ultimate secret of humanity is the coexistence of body and soul—distinct yet dwelling in equivalent physical spaces of being."

—The Author

This new paradigm does not negate Physicalism, the current paradigm of reality. Instead, it reinforces it by eliminating limitations caused by a narrow view confined to our universe. Multispace is an enhanced form of Physicalism, demonstrating that many items initially perceived as non-physical in the current paradigm are indeed physical in multispace reality.

26.1 EXPLAINING BODY AND SOUL

In physics, elementary particles are physical, even if they are explained to exist in quantum spaces and are dimensionless, meaning they do not exist in our physical universe. Elementary particles interact within our universe with each other and other energy fields through their fields that penetrate their (quantum) space boundaries.

In Multispace, the central challenge for physicalism, also known as the hard problem of consciousness or the mind-body dichotomy (see Subsection 25.3 on page 237), is resolved by explaining that souls are also physical and exist in their own frameworks. Similar to elementary particles that are physical in their spaces, our souls & minds penetrate their spaces and communicate with our bodies and brains.

Minds communicate both ways with their brains, as M. Gazzaniga, the famous neuroscientist and the founding Director of the Sage Center for the Study of the Mind at UC Santa Barbara (UCSB), said in 2015:

> *"99 percent of the brain's work goes on outside our conscious awareness [48]."*

While these innovative theories explore the possibility of souls & minds as physical quantum phenomena, they are not substantiated by current scientific evidence and reside at the fringes of interdisciplinary speculation.

Only now, the Multispace paradigm provides the correct explanation: our bodies and souls exist in their own equivalent and real spaces, not in illusory quantum spaces. The only difference is that, even if our bodies are built of invisible particles, we can see them, but not our souls, because—existing in their own spaces—they exist outside of this universe.

And we have proofs of their physicality.

26.2 SUBJECTIVE AGE

The subjective age refers to the age souls perceive themselves to be. If we possessed a non-physical mind (res cogitans), as René Descartes proposed, we would not experience the passage of time or the sensation of aging.

Yet, most people experience a difference between their felt (subjective) age and their chronological age, with subjective age often deviating from the actual number of years they've lived.

This difference is called *subjective age* and refers to how old a person feels, which can be younger, older, or the same as their chronological age. Studies show that adults typically report feeling younger than their chronological age, and this gap tends to grow with age—older adults often perceiving themselves as much younger than their birth date suggests [49].

For instance, nearly half of Canadians report feeling mentally younger than their chronological age, while about one in five feel older. Conversely, younger individuals often feel older due to adult responsibilities.

Subjective age is a universal concept. Most individuals feel younger than their actual age, but the extent of this gap and its implications differ based on culture, life experiences, and personal circumstances. For example, Americans report a larger subjective age gap than those from Germany or China.

* * *

Part X

CONVERGETICS MISSION

27 CONVERGETICS' MISSION

27.1 ELECTROMAGNETIC TRAVEL

A significant application of section 11.4 "Most Fundamental Cruise Control (MFCC)" on page 114 is electromagnetic travel with speeds at least 1% of the speed of light.

Electromagnetic travel requires the development of two technologies: (1) separating the vessel's space from the universe's space, and (2) propelling the vessel to "ride" on the universe's space.

27.2 GLOBAL EXPANSION SURVEY CENTER

The Convergetics Survey Center's mission is to monitor regions susceptible to faster expansion. It aims to measure crustal tensions and identify areas prone to fractures producing earthquakes and other natural cataclysms.

Ideally, this survey will be able to forecast close to 100% of earthquakes.

27.3 MULTISPACE PHYSICS

The present document establishes the foundations of Multispace Physics, which is currently in development.

27.4 ESCAPING EARTH

Presently, we do not know the exact future rate of Earth's expanding, warming, and shaking. The Thousand Year Project aims to find a new home for humanity before it becomes uninhabitable. Is migrating from Earth to Mars or other planet feasible, or is it progressing too slowly to save everyone?

27.5 NUTCRACKER PROJECT

The present nuclear fission technology splits heavy atoms into two or more lighter nuclei, with a significant release of energy in the form of heat and radiation.

The matter we observe was not created here but during inter-universe transfers through twisting processes during successive leaks from one universe to another.

This project aims untwisting elementary particles, extracting significant energy from thin air.

To release the energy trapped inside particles, one must reverse the process, must "untwist particles." This requires conditions we do not yet understand and are currently unobtainable with today's technology.

* * *

Part XI

APPENDIX

I SAMUEL WARREN CAREY

The problem is, what triggers the [expansion] process. As a geologist, I insist that the Earth has expanded, and leave it as a cosmological problem of the whole universe. Hence I do not see it as a problem specifically for the Earth, or for the solar system, but for the Universe and Cosmos.
—Samuel Warren Carey [22]

Samuel Warren Carey (1911 - 2002), a leading proponent of the growing Earth theory, was an Australian geologist and professor at the University of Tasmania. Initially, Carey supported the theory of continental drift. His research on plate tectonics reconstructions led him in the 1950s to develop the *Expanding Earth hypothesis*.

He demonstrated that *subduction could not balance the sea-floor spreading at oceanic ridges, presenting unresolved paradoxes that continue to challenge plate tectonics [21].*

I.1 CAREY'S THEORY

Carey proposed that plate movement originates from a continuous creation process within the Earth's interior, with new matter being added from space through accretion.

However, Carey couldn't explain the formation of matter inside the Earth and didn't see it as a significant issue. He believed the question was more fundamental:

> *"The problem is, what triggers the [expansion] process. As a geologist, I insist that the Earth has expanded, and leave it as a cosmological problem of the whole universe. Hence I do not see it as a problem specifically for the Earth, or for the solar system, but for the Universe and Cosmos [21]. "*

As discussed in previous chapters, Carey's intuition was correct, as the expansion process is indeed a fundamental physics and cosmology issue.

Carey also suggested that initially, the Earth grew through logarithmic mass growth, comparing its evolution to *the size spectrum of the other bodies of the solar system [22]:*

Without linking Carey's Earth and Cosmos theory to Minkowski's physics theory, we will never fully understand reality.

Refer to the following excerpt from the book *Earth Universe Cosmos*, by S. Warren Carey, Chapter 7: Earth Expansion [23]

1.2 EARTH UNIVERSE COSMOS

The following is an excerpt of the book *Earth Universe Cosmos*, by S. Warren Carey, Chapter 7: Earth Expansion [23]

I.2.1 Mass growth

I have already indicated, that if the Earth had formerly half its radius, mass must also have been much less, for otherwise the gravity acceleration at the surface would be four times its present value, which would certainly have been apparent from several phenomena.

Nernst (1920), followed by Hilgenberg (1933) and Egyed (1956), believed that the aether was a pool of energy which supplied the needed mass. Halm (1935) thought that Earth expansion related to the Hubble expansion of the Universe and that the radius of the hydrogen atom increased. Dirac (1937) studied large dimensionless cosmological numbers and argued that the gravitational constant has changed with time. He was later followed by Jordan (1952). Keindl (1940) and others have postulated an early metastable superdense Earth's core that expanded to present density.

For my part, I point out that no *free lunch* is involved because mass and energy are opposites (see Chapter 3), that E=mc2, and that mass and potential energy appear from nothing like Siamese twins, or like a bank loan and its complementary debt, that one cannot exist without the other, that they appear together from nothing and may vanish together to nothing, which could happen anywhere. The problem is, what triggers the process. *As a geologist, I insist that the Earth has expanded, and leave it as a cosmological problem of the whole universe.*

Hence I do not see it as a problem specifically for the Earth, or for the solar system, but for the Universe and Cosmos. Nevertheless it is useful to examine the relationship of Earth to its fellow travellers in the Solar system:

There we find a complete series of sizes, from irregular rocks to massive planets. Could this sequence recapitulate the evolution of the Earth, improbable as it seems, as a fertilized ovum recapitulates the evolution of the human race?

Let us now review the status of the Earth against the size spectrum of the other bodies of the solar system, assuming that it grew only by logarithmic mass growth.

At 100 km radius, such a proto-Earth would have been still an irregular rock with insufficient gravitation to command a spherical shape. Compare such a proto-Earth then with Hektor (73x150 km radius), Hestia (67), Himalia (85), Epimethus (70x40x50), and Puck (85). None are spherical, and all show impact craters, some large like the 50 km Bogle Crater on Puek.

By 200 km radius, proto-Earth would have been beginning to show some concern about her figure. Compare her with Phoebe (110 km radius), Juno (144x115), Amalthea (75x82x135), Europa (146), Neteid (300), Proteus (208), and Hyperion (205x30). None of these are yet spherical, but they are starting to show a tendency that way, and all have impact craters.

A radius more than 200 km, and mass 10^{19} *kg* would have been a critical transition for a proto-Earth. It would have become nearly spherical, and from this time show surface expansion cracks, and the first signs of volcanism, and hence a molten interior. Compare Mimas (197 km radius), now spherical; Miranda (242), spherical with an intense global extension system, and Enceladus (251), spherical with global tensional fractures and volcanic resurfacing.

Above 500 km radius, and mass 10^{20} *kg*, all bodies have spherical shape, pan-global extension fractures, repeated volcanic resurfacing, and the beginnings of a tenuous atmosphere from current out-gassing. Compare Tethys (530 km radius), Dione (560), Ariel (560), Umbriel (596), Iapetus (730), Rhea (765), Oberon (773), Europa (781), and Titania (805).

From the above comparison with much smaller bodies, the Earth should certainly then show volcanic activity. Surface gravity would be about the same as now, and a substantial atmosphere would be retained. According to Chin and Stothers (1975) climatic conditions and glaciation would not be improbable, even if the Sun and Earth were much smaller then.

Above 1000 km radius, and 10^{22} *kg mass*, all bodies are spherical, have extensive volcanism (some currently active) and resurfacing, a thin atmosphere, and the beginnings of a magnetic field. Compare Pluto (1150km radius), Moon (1738), Triton (1750), and Io (1816). But in this size range, some are currently active, and the still smaller ones show volcanism and resurfacing. But all should have frozen solid long ago. *So whence the heat?*

Stanton Pease actually predicted, a few days before Voyager I reached Jupiter, that Io's interior must be very hot or even molten because of tidal heating. A single satellite would quickly settle down in equilibrium resonance with its primary, which would end tidal heating. But Jupiter has three other, large satellites, Europa, Ganymede, and Callisto, which are in resonance with each other and with Jupiter, so that none of them quite settle down, and their Jupiter tidal bulges drag back and forth according to the periods of their conjunctions. Except near conjunctions, the tidal friction would be negligible.

Thus Europa is in resonance with Jupiter but completes two orbits to every one of Ganymede, and Io, also in synchronous resonance with Jupiter, completes two orbits for every one of Europa, hence an oscillating tidal bulge.

Marcialis and Greenburg have suggested similar heat sources for small satellites in chaotic rotation, with large inclinations. I doubt the validity of these proposals, because the energy expended each heating cycle alters orbits and periods, and there has been ample time for equilibrium adjustments to have been reached, just as Moon points her bulge permanently towards Earth (so there is no tidal heating on Moon), and Deimos and Phobos both point their tidal bulges permanently toward Mars. *All bodies in these size ranges exhibit or have exhibited strong volcanism, not just the few locked into tidal friction.*

Above 2,000 km radius, and mass 10^{23} *kg,* there is usually a substantial atmosphere, and a magnetosphere may appear for the first time, suggesting global differentiation with a convecting fluid zone. Compare Callisto (2,410 km radius), Mercury (2,433), Titan (2,575), Ganymede (2,631), Mars (3,380).

From this analysis, it is clear that it is possible to point to a continuous suite of models to match a waxing Earth at every stage from an irregular rock right through to the present mature globe. But at the end of the Archean (half her present age), Earth had half her present radius (say 3000 km). Even the preceding 2,000 million years is insufficient time [for Earth to grow to it's present size], because growth during the flat part of the logarithmic curve growth would be much too slow. *Accretion during the initial stage must have given Earth a substantial start.*

I.2.2 Synthesis

Expansion of the Earth has been most rapid during the last twentieth of geologic time! Certainly diapiric orogenesis is apparent in the oldest rocks we can study. Quite small bodies of the solar system show tensional fracture patterns. But the rate of expansion has accelerated. Figure 111 is a sketch of the whole Earth at the end of the Paleozoic. It wraps around so that the Americas adjoin Africa and Greenland. Neither the Pacific nor Atlantic Oceans existed at this time. All ocean floor that ever was, still is!

* * *

BIBLIOGRAPHY

[1] Philip W. Livermore et al. "Electromagnetically driven westward drift and inner-core superrotation in Earth's core." In: *PNAS* 110.40 (Oct. 2013), pp. 15914–15918. DOI: 10.1073/pnas.1307825110. URL: https://www.pnas.org/content/pnas/110/40/15914.full.pdf.

[2] Sandrine Codis et al. "Galaxy orientation with the cosmic web across cosmic time." In: *arXiv* (2018). DOI: 10.48550/arXiv.1809.06212. URL: https://arxiv.org/abs/1809.06212.

[3] Matsuura et al. *The dark lane of the planetary nebula NGC 6302*. 2005. DOI: 10.1111/j.1365-2966.2005.08903.x. URL: https://academic.oup.com/mnras/article/359/1/383/984080.

[4] NASA ESA et al. *Carina Cliffs from the Webb Space Telescope*. 2022. URL: https://apod.nasa.gov/apod/ap220905.html.

[5] Leopold Infeld Albert Einstein. *The Evolution of Physics: The Growth of Ideas from Early Concepts to Relativity and Quanta*. Ed. by David (2002) Harrison. Complementarity and the Copenhagen Interpretation of Quantum Mechanics. Cambridge University Press., 1938.

[6] Gene Alexandrescu. "Breaking the Cosmic Code: Quantum Spaces Are Real, Not Imaginary Structures." In: *Convergetics Research Center* (2023). URL: https://convergetics.org/2023/07/the-ultimate-

secret-of-the-universe-quantum-spaces-are-real-not-imaginary-spaces/.

[7] Gene Alexandrescu. "New Multiverse Paradigm of Physical Reality: The Theory of Decaying Universes." In: *Convergetics Research Center* (2023). URL: https://bit.ly/482RkPG.

[8] Clabon W. Allen and Arthur N. Cox. *Allen's Astrophysical Quantities*. Ed. by Arthur N. Cox. 4th. Springer, 2000. ISBN: 978-0-387-98746-0.

[9] Giorgio T. Bagni. *The Role of the History of Mathematics in Mathematics Education: Reflections and Examples*. Department of Mathematics. University of Bologna. URL: https://www.fmd.uni-osnabrueck.de/ebooks/erme/cerme1-proceedings/papers_vol2/g7_bagni.pdf.

[10] Harry Baker. "Earth's rotating inner core is starting to slow down— and it could alter the length of our days." In: *Live Science* (June 2024). URL: https://www.livescience.com/planet-earth/geology/earths-rotating-inner-core-is-starting-to-slow-down-and-it-could-alter-the-length-of-our-days.

[11] H.M. Basu S.; Antia. "Helioseismology and Solar Abundances." In: *Physics Reports* 457.5-6 (2008), pp. 217–283. DOI: doi:10.1016/j.physrep.2007.12.002. URL: arXiv:0711.4590.

[12] Friedrich Beck and John C. Eccles. "Quantum aspects of brain activity and the role of consciousness." In: *Proceedings of the National Academy of Sciences* 89.23 (1992), pp. 11357–11361.

[13] Iris Berent. "The "Hard Problem of Consciousness" Arises from Human Psychology." In: *Open Mind* 7 (2023), pp. 564–587.

[14] Paul Bloom. *Descartes' baby: How the science of child development explains what makes us human*. Basic Books, 2004.

[15] David Bohm. *Wholeness and the Implicate Order*. London: Routledge, 1980.

[16] Max Born and Albert Einstein. "The Born-Einstein Letters: Correspondence between Albert Einstein and Max and Hedwig Born from 1916-1955." In: (1971).

[17] Encyclopaedia Britannica. *Augustin-Louis Cauchy*. 2023. URL: https://www.britannica.com/biography/Augustin-Louis-Baron-Cauchy.

[18] C. D. Broad. *The Mind and Its Place in Nature*. Routledge & Kegan Paul, 1925.

[19] Ambros Brucker and Gerhard Bruschke. *Atlas of the World*. The Reader's Digest Association, Inc., 1990. ISBN: 2-7098-0317-8.

[20] B. Saxton C. Brogan. "HL Tau: Birth of Planets Revealed in Astonishing Detail." In: *National Radio Astronomy Observatory* (2014). URL: https://public.nrao.edu/gallery/hl-tau-birth-of-planets-revealed-in-astonishing-detail-2/.

[21] *The expanding earth — an essay review* 11 (1975). DOI: 10.1016/0012-8252(75)90097-5.

[22] S. Warren Carey. *Earth, Universe, Cosmos*. Vol. Geology Department. University of Tasmania, 1996. URL: https://books.google.ca/books/about/Earth_Universe_Cosmos.html?id=HE2ZGQAACAAJ&redir_esc=y.

[23] Samuel Warren Carey. *The Expanding Earth*. Vol. Earth Universe Cosmos. Chapter 7. URL: https://geocities.restorativland.org/CapeCanaveral/Launchpad/8098/EARTHEXP.htm.

[24] Susan Carey. "Where Our Number Concepts Come From." In: *Journal of Philosophy* 106.3 (2009), pp. 113–135.

[25] David J Chalmers. "Facing up to the problem of consciousness." In: *Journal of consciousness studies* 2.3 (1995), pp. 200–219.

[26] David J. Chalmers. "Strong and Weak Emergence." In: *The Re-Emergence of Emergence*. Ed. by Philip Clayton and Paul Davies. Oxford University Press, 2006, pp. 244–254.

[27] David John Chalmers. *The conscious mind: In search of a fundamental theory*. Oxford university press, 1996.

[28] Patrick Cornille and Akhlesh Lakhtakia. *Essays on the Formal Aspects of Electromagnetic Theory*. Ed. by Akhlesh Lakhtakia. 1st. Singapore: World Scientific, 1993, p. 796. ISBN: 9789810208547.

[29] Thibault Damour. "What is missing from Minkowski's "Raum und Zeit" lecture." In: *Arxiv* (2008). DOI: 10.48550/arXiv.0807.1300. URL: https://arxiv.org/abs/0807.1300v1.

[30] Donald Davidson. "Mental Events." In: *Experience and Theory*. Ed. by Lawrence Foster and J. W. Swanson. University of Massachusetts Press, 1970, pp. 79–101.

[31] W. F. G. Davies. "Maxwell's Original Equations." In: *The Mathematical Intelligencer* 38.2 (2016), pp. 6–14.

[32] Daniel Clement Dennett. *Consciousness explained*. Little, Brown, 1991.

[33] René Descartes. *Meditations on First Philosophy*. Hackett Publishing, 1993.

[34] Work by Descartes. "La Geometrie." In: *Britannica* (). URL: https://www.britannica.com/topic/La-Geometrie.

[35] Claudia Dreifus. "A CONVERSATION: With Sir Martin Rees; Tracing Evolution of Cosmos From Its Simplest Elements." In: *The Times and The Sunday Times* (2010). URL: https://www.nytimes.com/1998/04/28/science/conversation-with-sir-martin-rees-tracing-evolution-cosmos-its-simplest-elements.html.

[36] Editors of EarthSky. "Do all planets orbit in a flat plane around their suns?" In: *EarthSky* (Jan. 2019). URL: https://earthsky.org/space/planets-single-plane/.

[37] A.S. Eddington. "The Relativity of Time." In: *Nature 106* (1921), pp. 802–804.

[38] A. Einstein, B. Podolsky, and N. Rosen. "Can Quantum-Mechanical Description of Physical Reality Be Considered Complete?" In: *Physical Review* 47.10 (1935), pp. 777–780. DOI: 10.1103/PhysRev.47.777.

[39] Albert Einstein. "Zur Elektrodynamik bewegter Körper." In: *Annalen der Physik* 17 (1905), p. 891. URL: https://www.fourmilab.ch/etexts/einstein/specrel/www/.

[40] Albert Einstein. *Relativity: The Special and General Theory*. Methuen & Co Ltd, 1916. URL: https://www.gutenberg.org/ebooks/30155.

[41] Encyclopaedia Britannica. *Wave-particle duality*. 2025. URL: https://www.britannica.com/science/wave-particle-duality.

[42] Leonhard Euler. *Elements of Algebra*. Longman, Orme and Co., 1822. URL: https://openlibrary.org/books/OL23297090M/Elements_of_algebra.

[43] Moses Fayngold. *Special Relativity and How it Works*. Vol. p. 497. John Wiley and Sons, 2008. ISBN: ISBN 978-3-527-40607-4.

[44] James A. Foley. "Earth's Inner Core Spins Faster than Earth Itself." In: *NATURE WORLD NEWS* (Sept. 2013). URL: https://www.natureworldnews.com/articles/4014/20130917/earths-inner-core-spins-faster-earth-itself.htm.

[45] "Freeman Dyson." In: *Good Reads* (). URL: https://www.goodreads.com/author/quotes/5569343.Freeman_Dyson.

[46] Carl Friedrich Gauss. "Theoria Residiorum Biquadraticorum, Commentatio secunda." In: *Goettingen* Werke.Bd. 2 (1863), p. 177. URL: https://www.gutenberg.org/files/44730/44730-h/44730-h.htm.

[47] Yvon Gauthier. "Hermann Minkowski: From Geometry of Numbers to Physical Geometry." In: *Annalen der Physik* 17 (), pp 247–257.

[48] Michael Gazzaniga. *How does brain science affect our responsibility for actions?* 2015. URL: https://www.newscientist.com/article/mg22730303-500-how-does-brain-science-affect-our-responsibility-for-actions/.

[49] Sarah Grainger. *Why You Really Are Only as Old as You Feel*. Ph.D. New York: Psychology Today, 2025. URL: https://www.psychologytoday.com/ca/blog/aging-gracefully-the-scientific-scoop/202507/youre-only-as-old-as-you-feel.

[50] I. Grattan-Guinness and A. Walsh. *Benjamin Peirce*. 2001. URL: https://plato.stanford.edu/entries/peirce-benjamin/.

[51] Walter Greiner. *Quantum Mechanics: An Introduction*. Springer, 2001. ISBN: 978-3-540-67458-0.

[52] et al. H. Goenner. "The Expanding Worlds of General Relativity." In: *Einstein Studies* 7 (1999), pp. 45–86.

[53] Stuart Hameroff and Roger Penrose. "Consciousness in the universe: A review of the 'Orch OR' theory." In: *Physics of Life Reviews* 11.1 (2014), pp. 39–78.

[54] William Rowan Hamilton. "On Conjugate Functions, or Algebraic Couples," in: *British Association Report* (1834), pp. 519–523. URL: https://www.maths.tcd.ie/pub/HistMath/People/Hamilton/BACouples/BARep34B.pdf.

[55] William Rowan Hamilton. *Theory of Conjugate Functions, or Algebraic Couples; With a Preliminary and Elementary Essay on Algebra as The Science of Pure Time*. Vol. vol. 17, part 1. Transactions of the Royal Irish Academy, 1837. URL: https://www.maths.tcd.ie/pub/HistMath/People/Hamilton/PureTime/PureTime.pdf.

[56] William Rowan Hamilton. "Theory of Conjugate Functions, or Algebraic Couples; with a Preliminary and Elementary Essay on Algebra as the Science of Pure Time." In: *Transactions of the Royal Irish Academy* 17 (1837). Ed. by David R. Wilkins (online edition), pp. 293–422. URL: https://www.maths.tcd.ie/pub/HistMath/People/Hamilton/PureTime/PureTime.pdf.

[57] Yu He. "Superionic iron alloys and their seismic velocities in Earth's inner core." In: *Nature* 602 (Feb. 2022), pp. 258–262. DOI: 10.1038/s41586-021-04361-x. URL: https://www.nature.com/articles/s41586-021-04361-x.epdf.

[58] Werner Heisenberg. *There can be no doubt that consciousness does not occur in physics and chemistry, and I cannot see how it could possibly result from quantum mechanics*. Quoted in philosophical discussions on the limits of physical science and quantum mechanics concerning consciousness. The remark is attributed to Heisenberg and cited in several sources dealing with the philosophy of physics. 1955. URL: https://www.informationphilosopher.com/solutions/scientists/heisenberg/.

[59] "Imaginary Numbers." In: *The Proceedings of the Friesian School, Imaginary Numbers* (). URL: https://www.friesian.com/imagine.htm.

[60] Alexander Khot. *Observer's Mathematics Approach to Quantum Mechanics*. Accessed September 23, 2025. Luleå, Sweden: Conference on Mathematical Foundations of Quantum Mechanics, 2013. URL: http://staff.www.ltu.se/~norbert/JNMP-Conference-2013/khots.pdf.

[61] Jaegwon Kim. *Supervenience and Mind: Selected Philosophical Essays*. Cambridge University Press, 1993.

[62] Christof Koch and Klaus Hepp. "Quantum mechanics in the brain." In: *Nature* 440.7084 (2006), pp. 611–612.

[63] IDIA Visualisation Laboratory. *Astronomers reveal cosmic ribbon around rare galaxy*. 2023. URL: https://www.youtube.com/watch?v=dkMr-D2719w.

[64] Joseph Levine. "Materialism and qualia: the explanatory gap." In: *Pacific philosophical quarterly* 64.4 (1983), pp. 354–361.

[65] Alex Lobel. *The asymmetric Milky Way in motion*. 2022. URL: https://www.esa.int/ESA_Multimedia/

Videos/2022/06/The_asymmetric_Milky_Way_in_motion#.

[66] Pertti Lounesto. *Clifford Algebras and Spinors*. 2nd ed. Vol. 286. London Mathematical Society Lecture Note Series. Cambridge, UK: Cambridge University Press, 2001. ISBN: 978-0521005517. DOI: 10.1017/CBO9780511526022.

[67] Scalera G. Roberto Mantovani. "Scalera G. Roberto Mantovani and his ideas on the expanding Earth." In: *Ann. Geophys. [Internet]* (2009). URL: https://www.annalsofgeophysics.eu/index.php/annals/article/view/4622.

[68] James Maxlow. *Global Expansion Tectonics*. Oct. 2015. URL: https://www.jamesmaxlow.com/.

[69] James C. Maxwell. *A dynamical theory of the electromagnetic field*. Philosophical Transactions of the Royal Society of London, 1865. DOI: 10.1098/rstl.1865.0008.

[70] Hermann Minkowski. "Space and Time." In: *The Principle of Relativity: A Collection of Original Memoirs on the Special and General Theory of Relativity*. Ed. by Albert Einstein, H. A. Lorentz, Hermann Minkowski, and Hermann Weyl. English translation of the 1908 address "Raum und Zeit". New York: Dover Publications, 1923, pp. 75–91.

[71] Hermann Minkowski. "Space and Time & Raum und Zeit." In: *Fundamental Theories of Physics* Minkowski Spacetime: A Hundred Years Later (2010). URL: https://minkowskiinstitute.org/mip/MinkowskiFreemiumMIP2012.pdf.

[72] David Morin. "Frame of reference." In: *Introduction to Classical Mechanics* (2008), pp. 525–566. URL:

https://www.people.fas.harvard.edu/~djmorin/Relativity%20Chap%2010.pdf.

[73] Paul J. Nahin. *Dr. Euler's Fabulous Formula: Cures Many Mathematical Ills*. Princeton University Press, 2017. ISBN: 9780691175911.

[74] Rod Nave. "Time Dilation/Length Contraction." In: *HyperPhysics* (2025). Educational resource on special relativity and Lorentz transformations. HyperPhysics is authored by Dr. Rod Nave and maintained by Georgia State University. URL: https://hyperphysics.phy-astr.gsu.edu/hbase/relativ/tdil.html (visited on 09/23/2025).

[75] Graham Nerlich. *What Spacetime Explains: Metaphysical Essays on Space and Time*. Illustrated, reprint. Cambridge, UK: Cambridge University Press, Aug. 1994, p. 283. ISBN: 9780521452618.

[76] William L. Newman. *Geologic time: The age of the Earth*. USGS Unnumbered Series. USGS Publications Warehouse, 1977. DOI: 10.3133/70039222. URL: https://www.usgs.gov/publications/geologic-time-age-earth.

[77] John D. Norton. "General covariance and the foundations of general relativity: eight decades of dispute." In: *Rep. Prog. Phys.* 56 (1993), pp. 791–858. URL: https://sites.pitt.edu/~jdnorton/papers/decades.pdf.

[78] John D. Norton. "Discovering the Relativity of Simultaneity. How did Einstein take "The Step"?" In: *University of Pittsburgh* (). URL: https://sites.pitt.edu/~jdnorton/Goodies/rel_of_sim/index.html.

[79] "Notebooks of Einstein and Minkowski." In: *Jewish National and University Library* Box 9.7 (1904). URL: https://echo.mpiwg-berlin.mpg.de/content/modernphysics/jnul.

[80] Merriam-Webster Online. "manifold." In: *Merriam-Webster Online* (). URL: https://www.merriam-webster.com/dictionary/manifold.

[81] H.G. Owen. *Atlas of Continental Displacement: 200 Million Years to the Present*. Cambridge University Press, 1983. ISBN: 9780521258173. DOI: ISBN-10:0521258170.

[82] Hugh Owen. "The Earth Is Expanding and We Don't Know Why." In: *New Scientist* (Nov. 1984), p. 27. URL: http://bit.do/HughOwen84.

[83] Jason Palmer. "Earth's core far hotter than thought Published." In: *BBC News* (Apr. 2013). URL: https://www.bbc.com/news/science-environment-22297915.

[84] Kasha Patel. "Scientists discover yet another change to Earth's inner core: Its shape." In: *The Washington Post* (Feb. 2025). URL: https://www.washingtonpost.com/climate-environment/2025/02/10/earth-inner-core-shape-study/.

[85] Roger Penrose. *Roger Penrose: Collected Works: Volume 1: 1953-1967*. Oxford University Press, 2010.

[86] Sir Roger Penrose. *On the Origins of Twistor Theory, Gravitation and Geometry*. Naples, 1987. URL: https://users.ox.ac.uk/~tweb/00001/index.shtml.

[87] Doctor Pete. "Defining Complex Numbers." In: *The Math Forum* (). URL: httpa://mathforum.org/library/drmath/view/53809.html.

[88] Vesselin Petkov. "Physical Laws and Worldlines in Minkowski Spacetime." In: *Fundamental Theories of Physics* Minkowski Spacetime: A Hundred Years Later (2010), p. 285. DOI: 10.1007/978-90-481-3475-512.

[89] Vesselin Petkov. "Preface." In: *Fundamental Theories of Physics* Minkowski Spacetime: A Hundred Years Later (2010). DOI: 10.1007/978-90-481-3475-512. URL: https://minkowskiinstitute.org/mip/Minkowski-FreemiumMIP2012.pdf.

[90] U. T. Place. "Is Consciousness a Brain Process?" In: *British Journal of Psychology* 47.1 (1956), pp. 44–50.

[91] Dean Radin. *Entangled Minds: Extrasensory Experiences in a Quantum Reality*. New York: Paraview Pocket Books, 2006.

[92] Martin Rees. *The multiverse: our universe is suspiciously unlikely to exist—unless it is one of many*. 2023. URL: https://theconversation.com/the-multiverse-our-universe-is-suspiciously-unlikely-to-exist-unless-it-is-one-of-many-200585.

[93] Martin J. Rees. *Our Cosmic Habitat*. Princeton, NJ: Princeton University Press, 2001. ISBN: 978-0691095509.

[94] ESA Media Relations. *Gaia data release 3*. 2022. URL: https://www.cosmos.esa.int/web/gaia/data-release-3.

[95] ESA Media Relations. *Gaia data release 3*. 2022. URL: https://www.gaia.ac.uk/data/gaia-data-release-3.

[96] ESA Media Relations. *Gaia data release 3: exploring our multi-dimensional Milky Way*. 2022. URL: https://www.esa.int/ESA_Multimedia/Videos/2022/

06/Gaia_data_release_3_exploring_our_multi-dimensional_Milky_Way.

[97] ESA Media Relations. *Gaia overview*. 2022. URL: https://www.esa.int/Science_Exploration/Space_Science/Gaia/Gaia_overview.

[98] ESA Media Relations. *The Orion Nebula view with the NIRCam*. 2022. URL: https://www.esa.int/ESA_Multimedia/Images/2023/09/Orion_Nebula_in_NIRCam_long-wavelength_channel.

[99] ESA Media Relations. *You are here*. 2022. URL: https://www.esa.int/Science_Exploration/Space_Science/Gaia/Gaia_sees_strange_stars_in_most_detailed_Milky_Way_survey_to_date.

[100] Christiane Rousseau. *How Inge Lehmann Discovered the Inner Core of the Earth*. Vol. 44. The College Mathematics Journal 5. The Mathematical Association Of America, 2013. URL: https://www.maa.org/sites/default/files/pdf/upload_library/2/Rousseau-CMJ-2014.pdf.

[101] Jeffrey M. Schwartz, Henry P. Stapp, and Mario Beauregard. "Quantum physics in neuroscience and psychology: A neurophysical model of mind–brain interaction." In: *Philosophical Transactions of the Royal Society B: Biological Sciences* 360.1458 (2005), pp. 1309–1327.

[102] Anna Sfard. "On the Dual Nature of Mathematical Conceptions: Reflections on Processes and Objects as Different Sides of the Same Coin." In: *Educational Studies in Mathematics* 22.1 (1991), pp. 1–36. DOI: 10.1007/BF00302715.

[103] Florentin Smarandache. *Multispace & Multistructure: Neutrosophic Transdisciplinarity (100 Collected Papers of Sciences), Vol. IV*. Ed. by Florentin Smaran-

dache. Hanko, Finland: North-European Scientific Publishers, 2010. ISBN: 978-952-67349-2-7.

[104] J. J. C. Smart. "Sensations and Brain Processes." In: *Philosophical Review* 68.2 (1959), pp. 141–156.

[105] Precious Smith. "Scientists Discovered Unexpected 'Hidden World' in Earth's Inner Core!" In: *Nature World News* 28 (Oct. 2021). URL: https://www.natureworldnews.com/articles/48006/20211028/scientists-discovered-new-hidden-world-in-earths-inner-core.htm.

[106] John Stachel and al. *The Collected Papers of Albert Einstein*. Vol. 1: The Early Years, 1879—1902. Doc. 28. Princeton: Princeton University Press, 1987.

[107] Henry P. Stapp. *Mindful Universe: Quantum Mechanics and the Participating Observer*. Berlin: Springer, 2007.

[108] Daniel Stoljar. "Physicalism." In: (2024). Ed. by Edward N. Zalta and Uri Nodelman. URL: https://plato.stanford.edu/archives/spr2024/entries/physicalism/.

[109] P. Sudiro. "The Earth expansion theory and its transition from scientific hypothesis to pseudoscientific belief." In: *Hist. Geo Space. Sci.* 5.1 (2014). DOI: 10.5194/hgss-5-135-2014.

[110] J. L. Synge. *Relativity: the general theory*. Vol. IX. North-Holland, Amsterdam, 1960.

[111] Max Tegmark. "Importance of quantum decoherence in brain processes." In: *Physical Review E* 61.4 (2000), pp. 4194–4206.

[112] The American Heritage Dictionary Editors. *The American Heritage Dictionary of the English Language.* Fourth Edition. Accessed September 23, 2025. Boston, MA: Houghton Mifflin Harcourt, Sept. 2000. ISBN: 978-0395825174. URL: https://www.ahdictionary.com/word/search.html?q=imaginary.

[113] *The American Heritage Dictionary of the English Language,* 4th ed. Houghton Mifflin Harcourt; Thumb Indexed edition, 2000. DOI: ISBN_978-0395825174. URL: https://www.ahdictionary.com/word/search.html?q=real+number.

[114] "The Imaginary Unit i as an Operator." In: *Mathematics Stack Exchange* (). URL: https://math.stackexchange.com/questions/692888/the-imaginary-unit-i-as-an-operator.

[115] *The Tao of Physics: An Exploration of the Parallels between Modern Physics and Eastern Mysticism Paperback.* 5th ed. Shambhala, 2010. DOI: ISBN_978-1590308356. URL: https://www.amazon.ca/Tao-Physics-Exploration-Parallels-Mysticism/dp/1590308352/.

[116] Michael Tsamparlis. *Special Relativity: An Introduction with 200 Problems and Solutions.* 2nd ed. Undergraduate Lecture Notes in Physics. Cham, Switzerland: Springer, 2019. ISBN: 978-3-030-27346-0. DOI: 10.1007/978-3-030-27347-7. URL: https://www.springer.com/gp/book/9783030273460.

[117] Jack A. Tuszynski, ed. *The Emerging Physics of Consciousness.* Berlin: Springer, 2006.

[118] Scott Walter. "The Non-Euclidean Style of Minkowskian Relativity." In: *The Symbolic Universe* (1999), pp. 91–127. URL: https://www.univ-nancy2.fr/DepPhilo/walter/papers/nes.xml.

[119] Scott Walter. "Book review: Beyond Einstein's Velocity Addition Law." In: *Fundamental Theories of Physics 117* (2001). URL: https://www.academia.edu/en/43865238/Book_Review_Beyond_Einsteins_Velocity_Addition_Law_By_Abraham_A_Ungar_Fundamental_Theories_of_Physics_117_Kluwer_Academic_Dordrecht_The_Netherlands_2001_xlii_413_pp_138_00_hardcover_.

[120] Scott Walter. "Minkowski, Mathematicians, and the Mathematical Theory of Relativity." In: *Bullettin of the American Mathematical Society* 45.4 (2008), pp. 635–648. ISSN: 0273-0979. DOI: 10.1090/S0273-0979-08-01234-5.

[121] Steven Weinberg. *The Quantum Theory of Fields, Volume 1: Foundations*. Cambridge: Cambridge University Press, 1995.

[122] Galina Weinstein. "Max Born, Albert Einstein and Hermann Minkowski's Space-Time Formalism of Special Relativity." In: *Arxiv* (2012). URL: https://arxiv.org/pdf/1210.6929.

[123] Eric W. Weisstein. *Argand Diagram*. MathWorld. URL: https://mathworld.wolfram.com/ArgandDiagram.html.

[124] Tanja Woyke and Frederik Schulz. "Entities Inside One Another: A Matryoshka Doll in Biology?" In: *Perspectives in Biology* 7.2 (2019), pp. 123–127. URL: https://pmc.ncbi.nlm.nih.gov/articles/PMC7379638/.

[125] Holly Zell. *Solar Rotation Varies by Latitude*. Explains the Sun's differential rotation by latitude and the observational origins of this discovery. 2013. URL: https://www.nasa.gov/mission_pages/sunearth/science/solar-rotation.html (visited on 09/23/2025).

www.ingramcontent.com/pod-product-compliance
Lightning Source LLC
LaVergne TN
LVHW052255200926
848481LV00002B/378